Bee-Keeping
at
Buckfast Abbey

BEE-KEEPING
AT
BUCKFAST ABBEY

With a Section on Meadmaking

BY

BROTHER ADAM, O.B.E.

NORTHERN BEE BOOKS
Mytholmroyd : Hebden Bridge

British Library Cataloguing in Publication Data

Adam, *Brother*
Bee-keeping at Buckfast Abbey:
with a section on meadmaking. – 4th ed.
1. Bee culture – England.
I. Title
638'.1'0924   SF531.G7

ISBN 0-907908-37-3

First published 1975
Reprinted   1977
Reprinted   1980
Reprinted   1987

This fourth English edition published by Northern Bee Books,
Scout Bottom Farm Mytholmroyd, Hebden Bridge, West Yorkshire.
© May 1987

This book is in part based on 'Meine Betriebsweise', first published 1971
by the Ehrenwirth, Verlag, Munich, West Germany.

CONTENTS

PREFACE

PART I

GENERAL OBSERVATIONS

PART II

SEASONAL MANAGEMENT

PART III

BREEDING AND RAISING OF QUEENS

PREFACE

For many years I have been urged to write a book on our way of bee-keeping. However, more important undertakings, which would not brook any delay, have hitherto prevented me from giving my attention to the request indicated. Realising that it was a case of either now or never, I have made an effort to meet the widely expressed wish.

Bee-keeping at Buckfast can look back on a long tradition. Indeed, the bee boles in a part of an ancient enclosure wall seem to indicate that bees may have been kept at the Abbey before the dissolution in 1539. Anyway there were bees on the property when the monks returned to Buckfast in 1882. My own recollections go back to 1910, and my direct participation in our bee-keeping commenced just five years later, when in March 1915 I was appointed to assist Br. Columban, who had been in charge of our bee-keeping since 1895.

My introduction to bee-keeping could hardly have come at a less propitious moment. The Isle of Wight disease, or Acarine, was at its height, and in the autumn of 1915 the county bee-keeping officer predicted that by next spring we would have no bees left. However, of the 46 colonies, 16 survived. All the native colonies succumbed but those of Italian origin withstood the disease. The summer of 1916 was a favourable one and we were able to make good the losses of the previous winter, and in 1917 we increased the number of our colonies to 100. During the next two years we devoted all our efforts to helping the general restocking of the bee population by dispatching literally hundreds of colonies to all parts of the country. On 1st September, 1919, Br. Columban retired and I was given sole charge of bee-keeping at the Abbey.

I mention these points because I am convinced that it is almost impossible for anyone who did not experience at first-hand the problems and disheartening losses when Acarine raged through the apiaries of the country, nor to appreciate fully the immense strides made since 1920 in all practical and technical aspects of bee-keeping. This progress was achieved step by step and as a result of untold effort. The illustrations I have selected will in some measure convey an idea of the advancements made over the past sixty-five years.

This book is not a manual, but a general account of the beekeeping as carried out at Buckfast. As will be noted, every item of equipment, every manipulation, every aspect of management has been designed to achieve the best possible results, calling for a minimum of effort and time on our part. A due regard is simultaneously paid to the ways and instincts of the honeybee. Though she is endowed with a marvellous gift of adaptation, we cannot with impunity disregard her truly wonderful organisation and immutable instincts. Indeed, it is one of the foremost tasks of every bee-keeper to study and adjust himself to her ways, if he wishes to succeed in bee-keeping.

I have not dealt with the contribution made towards the advancement of bee-culture by the journeys of research and the evaluations of the various races we have carried out during the past twenty-five years. This aspect of our endeavours is fully covered in the book 'In Search of the Best Strains of Bees.'

I would like to take this opportunity to express my deep appreciation and gratitude to the Rev. Fr. Leo Smith, PhD, who very kindly undertook to check the manuscript and for suggestions put forward. I also wish to express my thanks to all the helpers who in the course of the past sixty years have given me their assistance in one way or another. Without their support the work accomplished would not have been possible.

BR. ADAM
Spring 1974

PREFACE TO THE FOURTH EDITION

This book has seemingly met a special need. Since its first appearance about twenty years ago it has been published in no less than five languages: English, French, German, Greek and Swedish.

Notwithstanding the various translations and editions issued, no revisions have at any time been called for. The special equipment and management described have stood the test of time and also confirmed the findings of the previous fifty years. But in the case of this edition a few particular aspects of our beekeeping apparently call for an emphasis. Their far-reaching importance has not always been fully appreciated.

Maximum yields of honey per colony, entailing a minimum of effort and time, have at all times constituted our primary objective. To achieve this end queens of the highest standard, physically and genetically, form the linchpin. That these endeavours over the years have met with a measure of success is confirmed by the results secured on a world wide basis. Even in Europe averages of up to 182 kg and an individual yield of 253 kg – equalling 558 lb – have been recorded. However we must at the same time accept that the best of strains, skill and equipment are of little avail when in seasons of adverse climatic conditions every source of nectar fails.

When all is said and done, success in beekeeping is, in its final analysis, determined by our ability to ensure that every colony is at all times in the best condition to make the most of a honey flow when one comes along.

Br. Adam
Spring, 1986

PART I
GENERAL OBSERVATIONS

There is something about bees that in all ages has taken men captive. Their extraordinary sense of order and precision, their ability to adapt themselves to anything and everything, their amazing versatility, these and many other characteristics provide an inexhaustible source of interest and delight for the professional bee-keeper with his 2,000 stocks as well as for the amateur with his few hives in a corner of the garden.

Success in bee-keeping is largely based on the application of some sound commonsense principles to seasonal management. In the second part of this book I will describe in detail our seasonal management which we have evolved in the course of the past fifty years. I will also endeavour to indicate why a particular measure was adopted by us, and the advantages and disadvantages of certain methods in common use among bee-keepers.

Clearly the approach of the commercial bee-keeper must be different from that of the amateur, yet there are certain basic principles in the management of bees which neither the one nor the other can neglect. Before describing therefore our seasonal management I should like to deal with these basic principles which are so often neglected in manuals of bee-keeping.

1. PRINCIPLES OF MANAGEMENT

There was a time, not so many years ago, when great value was placed on certain particular methods of management, based on a complete disregard of the truly marvellous organisation and wisely balanced interactions regulating the activities of a colony of honeybees. However, experience has shown that all such intrusions and lack of elementary considerations not only usually fail to achieve the intended results but in fact proved positively harmful to the well-being of a colony. Indeed, were it not for the most extraordinary ability of the honeybee to overcome and adjust itself to the most flagrant, wanton interference of many well-meaning custodians, success in the keeping of bees would prove even far more of a hazard than is actually the case. There is doubtless some truth in the assumption that in many instances a colony will produce a surplus *in spite of the bee-keeper.*

At Buckfast we endeavour, so far as is possible, to respect the inviolableness of the mainspring of the life of a colony, namely, the brood-nest. The 'spreading of brood', the removal of pollen-clogged combs to hasten the spring build-up, stimulative feeding, every unnecessary examination and disturbance are strictly banned and have no place in our management. Emphasis is placed, too, on the use of a minimum of smoke, no more than really called for. Indeed, the correct use of smoke for subjugating bees is an art.

Wherever bee-keeping is carried out with a remunerative end in view the time factor of every step or measure composing a scheme of management must necessarily play an important role. In fact the financial returns will largely be determined by a close scrutiny and assessment of the time each operation demands. Every device and simplification, resulting in a saving of time and energy, is of the utmost importance in commercial bee-keeping. The design of the various items of equipment, no less so the strain of bee used, will have a decided bearing on the time we have to spend on a colony in the course of a season. The cumulative effect of a particular action, carried out perhaps a hundred times a day during the busy part of the season, is a matter of vital concern to the professional bee-keeper.

Before going into details of the different considerations indicated, it will be appropriate to relate here an episode of one of my earliest bee-keeping experiences – an incident which had a far-reaching influence on all subsequent developments of our beekeeping. As is so often the case, seemingly minor events reveal information of major importance and great consequence.

When I was placed in charge of our bee-keeping department in 1919, it was at a point in time when the keeping of bees in this country was up against a series of momentous problems. The Isle of Wight disease, as it was then known, had swept away not only the old English native bee, but also in large measure the kind of bee-keeping as practised up to then. Bee-keeping had perforce to re-orientate and adjust itself to the new conditions. The old-time ideas and tenets, which before the demise of the old native bee had a measure of justification, no longer held good. The Italian bee, which now had usurped an almost unchallenged predominance, by reason of her high resistance to the disease that proved to be the doom of the native bee, could not be managed on the same lines as the former indigenous variety. As is so often the case, the old-time bee-keepers had great difficulty in accepting and adjusting themselves to the new circumstances forced on them. It was widely held at that time that a non-native bee, as well as foreign methods of bee-keeping, were quite unsuitable for the conditions prevailing in the British Isles. These views, held with the utmost conviction, were largely based on a misapprehension. On the other hand, even the most progressively minded beekeeper could not at that time visualise the developments the future had in store. For instance, in 1906 a bee-keeper in this neighbourhood secured from one colony, headed by a cross-queen, a surplus of 160 lb., then considered an all-time record. However, since 1920 we obtained one year an overall average of no less than 192 lb. of surplus per colony and individual yields exceeding 3 cwt. While high individual colony yields are relatively common it is the overall average returns, over a series of years, that count.

Up to 1920 our colonies were managed, as was the custom then, on one ten-frame British Standard brood chamber. We were aware that a brood chamber of this capacity was quite inadequate for the more prolific Italian bee and especially so when crossed. Thus restricted the Italians had no chance of attaining their maximum colony strength and maximum potential honey-gathering capability. Indeed, the severe restriction gave rise to undue swarming which in turn reduced still further the honey-gathering potential of such colonies. But tradition and the conventional views then current were decidedly against the use of two brood chambers.

In the autumn of 1920 we tentatively placed one colony on two brood chambers, furnished with approximately 40 lb. of stores. In the following spring this colony was far ahead of the colonies wintered on ten combs. The spring build-up of that colony exceeded our most optimistic expectations and by the time the orchards were in bloom it covered the two brood chambers and was ready for its first super. Apart from providing supers as required – of which it needed no less than six – this colony caused us no work throughout the season. By the end of July it occupied two brood chambers and six supers and towered like a lighthouse over the rest of the hives. It did in fact prove a lighthouse not just in a symbolical sense, but in a severely practical one. It showed up not only the direction and course our bee-keeping would have to take, but also indicated the rocks and sandbanks which all too often lead to shipwreck in modern bee-keeping. We cannot interfere at will in the highly organised life of a colony with impunity.

This colony was of course *the* perfect example, but its performance was by no means out of the way. That one colony gave a yield of 1½ cwt. without any trouble or time-consuming attention – or an additional 40 lb. in excess of the general average that season. Then we ask the question: What was the reason for this surprising success? There can be no doubt that the answer lies in the larger hive, or more acurately, in the unlimited breeding space provided by the use of two brood chambers and the provision of an abundance of winter stores.

These two factors together, the unlimited breeding space and the adequate supply of stores, enabled the development of the colony in the spring to proceed uninterruptedly without any kind of stimulative feeding. But ample breeding space and an abundance of stores would have been to no purpose without a first-class queen of proven fecundity and stock known for its honey-gathering abilities. In spite of the good queen this colony would not have made a pound more honey than the others in the apiary, had not the other factors been present too. On the other hand, we know from bitter experience that without a first-class queen all the other factors are of no avail, in fact they can be a disadvantage. Moreover, we know that a colony, given these essential conditions, will not only produce more honey, but produce it with a minimum of labour and time given to it on the part of the bee-keeper.

To sum up: this incident provides us with a scale of values, a list of priorities, for successful and economically sound bee-keeping. First, we must have a bee which is able to meet the demands of modern bee-keeping. Second, we need a type of hive with a brood chamber of sufficient size to enable a colony to attain its highest possible honey-gathering ability. Third, the bee-keeper must at all times heed the instincts and highly developed organisation of the bee.

2. THE BEE

There is no gainsaying that the strain of bees is the first and most important factor whether bees are kept for pleasure or profit. Moreover, any scheme of management will largely be determined by the kind of race and strain kept, so a brief indication of the particular type of bee kept by us is essential.

The Buckfast strain has been evolved from a cross between the leather-coloured Italian bee and the old native English variety. The original cross was formed about 60 years ago, that is before the native bee was eradicated by the Isle of Wight epidemic. Furthermore, the dark leather-coloured Italian, available at that time, differed in many ways to the strains now imported from Italy. In colour the Buckfast bee closely resembles the classic leather-coloured Italian of the Ligurian Alps, but we have at no time bred for uniformity in external characteristics, for such an aim can only be attained at the cost of performance. In any case, a high uniformity in external markings is hardly found in any race of bees, least of all in the Italian. Compared to the present-day Italian, the Buckfast bee is more industrious, more thrifty, less disposed to swarm, more resistant to disease, particularly Acarine. She collects less propolis than most strains, keeps restful in winter, but builds up rapidly at the appropriate time in spring and maintains a maximum effective colony strength throughout the summer, enabling her to take full advantage of a honey flow whenever one sets in. As for temper, she is unusually docile and will tolerate handling in unfavourable weather.

Of all the qualities a strain may possess, there is probably none more important than a disinclination to swarm. A strain may possess every desired characteristic, but a highly developed swarming instinct will effectively neutralise the qualities of

14

economic value. Swarming is clearly the bugbear of modern beekeeping.

While the pure Buckfast bee is very little disposed to swarm, it would be wrong to assume that it will not swarm in any circumstances. We shall in all likelihood never possess one that will manifest no disposition to swarm regardless of all possible circumstances. On the other hand, there are races and strains that will attempt to swarm irrespective of every preventive measure. Others will rarely swarm, often not at all. A commercial beekeeper, operating close to 2,000 colonies, held the view that our strain when properly managed swarmed so little as to render periodic examinations unremunerative. There is no doubt, swarming is no longer the problem it was at one time.

Though the pure Buckfast bee has a claim to an unusual honey-gathering ability we nevertheless rely in part on cross-bred stock for maximum returns. However, the pure strain forms the basis, mostly on the paternal side, in the case of almost every cross. As in most forms of life, more particularly in the honeybee, maximum returns are only possible in conjunction with hybrid vigour or heterosis. Indeed, it is only in cross-bred stock that the bee-keeper can hope to secure the full benefit of a highly developed pure strain. But due regard has to be paid to certain peculiarities in hybrids (more particularly in a first-cross) in the case of the honeybee, peculiarities which do not and cannot manifest themselves in other forms of life. These undesirable results of heterosis will be dealt with in the appropriate section of this book.

I have to draw a distinction between the hybrids I have here in mind and the generality of mongrel bees. In the case of hybrids and cross-bred stock under consideration the parentage is always known – anyway maternally – and is of pure origin. In mongrels the reverse is the case. Admittedly, mongrels usually possess the ability of surviving in the most unfavourable seasons without any attention on the part of their owner. But they are generally quite unreliable in all other respects, and almost invariably of an aggressive disposition, and given to swarming in and out of season. Occasionally a mongrel colony will manifest an exceptional honey-gathering ability but more often than not mongrels are completely unreliable.

It is often assumed that the dark or black bees, found at present in every part of the British Isles, represent the former English indigenous race of bee. This is a completely erroneous assumption. Indeed, I have never met anybody who knew the old native bee at first-hand who ventured to confirm such an assumption. The native bee had undoubtedly many extremely valuable characteristics, but equally so a great many serious defects and drawbacks. She was very bad tempered and very susceptible to brood diseases and would in any case not have been able to produce the crops we have secured since her demise.

The present-day dark bees originate from importations made in 1919 and subsequent years in an effort to restock the country. These importations came mainly from Holland and France and the bees brought in belonged to the same group of races as our former native variety. However, the latter differed in some characteristics from the importations, in spite of the close relation.

The great majority of the different sub-varieties of the western European race group have been put to the test in our apiaries – ranging from Finland in the extreme north to the most southerly point of the Iberian Peninsula – and we found that none of them would meet our present-day requirements. Indeed, the rapid decline in the popularity of the western European bee, and its progressive elimination in many countries, is a clear confirmation of our own findings.

While we fully appreciate the outstanding qualities of the western European dark bees – of which we kept as many as a 100 colonies at one time – we would never again consider giving them an extensive trial. Their undesirable traits far outweigh their good points.

One thing is self-evident: there is no race or strain that will meet the wishes of everyone. Moreover, there is no perfect or ideal bee. The choice will in every case mean a balancing of one set of advantages against another set of disadvantages, and an eventual adjustment to the particular idiosyncracies of the bee favoured.

3. THE HIVE

I now pass to make a few observations on the hive we are using and some indications of the reasons which prompted us to adopt this particular type, as for almost 50 years we used British Standard equipment before the change-over to the 12-frame Modified Dadant hive in 1930.

Needless to say, the type of hive used has a bearing on the results obtained in honey production. On the other hand, we must not lose sight of the fact that a modern hive is in many respects merely a tool in the hands of the bee-keeper. Bees are by nature extraordinarily undemanding and accommodating; a hollow tree, a cavity in a rock or wall having formed her normal habitation from time immemorial.

Perfection in a modern hive is not found in a complicated design nor in a multitude of gadgets, but on the contrary in an extreme simplicity of every detail. The more conveniently and more rapidly, and with the least effort the seasonal manipulations can be carried out, the more perfect the hive from the strictly practical point of view. It is indeed surprising how one can produce honey with best possible success by the use of the simplest of makeshift hives and equipment. There is, however, one really vital consideration; namely, the capacity of the brood chamber.

From 1920 to 1930 most of our colonies were accommodated in make-shift hives composed of 10-frame BS brood chambers, made of ½ in. timber; the floors and crown boards originated from disused packing cases; single sheets of Pluvex felt had to serve as roofs. To prevent an undue loss of warmth, a few newspapers were placed between the crown board and felt. Last thing in the autumn each hive was wrapped in an additional sheet of felt to keep them dry. I could never observe any noticeable difference in the amount of honey produced between these colonies and those in the dual-walled WBC hives. From this point of view the make-shift hives proved completely satisfactory.

With the spread of the Italian bee after the conclusion of the first world war, bee-keeping in this country experienced a new era of prosperity. This new prosperity was the result in a change of outlook and the adoption of more advanced methods of bee-keeping. It was now widely accepted that the more prolific Italian bee demanded a very much larger breeding space than afforded by a single 10-frame BS brood chamber. Also, that a ruthless simplification in the design and construction of hives and appliances was essential to render bee-keeping remunerative. The choice between single and double walled hives was no problem. It had necessarily to be in favour of the former. The needed capacity of breeding space, whether it should be provided in one large unit or two smaller ones, was less easy to decide. The majority of the more progressive bee-keepers favoured the Langstroth hive. But a single Langstroth brood chamber, though considerably larger than the British Standard, was still too small for a prolific queen. We did not wish to use two Langstroth brood chambers as is the common practice in America, and wherever this hive has been adopted. Indeed, I saw no real advantage when thus used between the Langstroth and the British Standard. Twenty combs had to be examined in either case. Moreover, as experience has amply demonstrated, the space between the two brood chambers tends to act as a barrier to the queen with the result that, though the breeding space is there, the actual area of brood rarely if ever equals that when no such barrier obstructs the movement of queens. Our final choice, therefore, fell on a brood chamber holding 12 Modified Dadant frames. A brood chamber of this size is square and measures 19⅞ in. x 19⅞ in. x 11⅞ in. in depth. The supers are of identical size but only 6 in. deep and hold ten wide shallow frames.

Our hive differs in a number of important features from the American design. Our bottom boards extend only an inch at the front and incorporate a slope from back to front, to prevent rain from entering and also to facilitate the clearance of debris by the bees. The upper part is rabbeted and a ⅜ in. tapered projection on the inside holds the brood chamber firmly in position. Unlike the American pattern, the bee space is provided under the frames in the brood chamber and supers. The Hoffman spacing never appealed to me and we have been using a form of hobnail, four to each brood frame, which we regard as the ideal method of spacing. Unfortunately, this particular type of hobnail is no longer obtainable and we now have to use 5⁄16 in. diameter screw eyes. Both hobnails and screw eyes provide a minimum point of contact and therefore facilitate the removal of the combs for inspection immensely. In the case of the supers, the spacing of the shallow frames is effected by a series of notches, which hold the lugs of the top-bars in position, doing away with any attachment to the frames. We have found these innovations most satisfactory from every point of view.

Whilst we lay great importance on simplicity, we value no less a durable and solid construction of every hive component, in particular that of the frames. Any attempt at economy in this direction is in my estimation a grave mistake. There is surely nothing more time-consuming and annoying than frames which do not retain their true shape or that fail to stand up to many years of usage. The same goes for hive parts which fail prematurely.

As already intimated, the capacity of the brood chamber is the only factor that has a bearing on the amount of honey a colony can gather. A brood chamber that constricts the potential laying ability of a queen will, *ipso facto,* frustrate the attainment of a maximum colony strength and of necessity, reduce in a corresponding measure the actual amount of honey such a colony would be capable of gathering. It will be readily appreciated that a restriction of the laying powers of a queen, accommodated for instance in a ten-frame BS brood chamber, must inevitably entail an equalisation of colony strength and honey-gathering ability. In such instances a potential maximum performance is rendered impossible and an objective evaluation for the purpose of selective breeding is lost at the same time. Furthermore, in cases of this kind an inordinate amounting of swarming will take place because of the inadequate breeding space, resulting in an additional distortion of the potential honey-gathering ability of individual colonies. Indeed, there is little doubt that many of the disappointments which have been recorded in the use of high performance strains, can be traced to a lack of an appreciation of essentials such as ample brood chamber space on the part of the bee-keeper.

I must confess, that in 1923, I was unaware of these vital and far-reaching considerations. My choice of a hive holding twelve MD frames in the brood chamber rested exclusively on technical consideration. Indeed, the leading authorities at the time predicted dire failure with a hive of this capacity, particularly on Dartmoor for the heather honey. However, I accepted certain possible disadvantages, in view of the very substantial benefits a single large brood chamber and one compact brood-nest provided. However, the results secured quickly demonstrated that these good people failed to visualise the possibilities awaiting us, for their assumptions were based on the limitations set by the old-English bee. Had we placed this bee in a twelve-frame MD hive their views would doubtless have been fully vindicated.

It will be readily appreciated that a change-over of this kind would not have been justified without tentative tests carried out over a period of years. The most convincing arguments and conclusions concerning all practical aspects of bee-keeping require substantiation by concrete results. In order to secure positive comparisons of the kind needed, we transferred in the summer of 1924, half the number of colonies in each out-apiary into twelve-frame MD hives. The other twenty colonies in each apiary remained on BS combs and on two brood chambers. The summer of 1924 was a poor one, but we managed to carry out the change-over without undue difficulty. The following summer proved outstandingly good. The comparative results achieved between the colonies on BS combs and those in the MD hives were indeed startling. The colonies in the MD hives fulfilled our most optimistic expectations, not only in regard to every practical consideration, but foremost in a very substantial difference in the amount of surplus produced. On the Moor the yields were approximately double that of the colonies in the smaller hives, a result confirmed in the succeeding years. These comparative tests, involving 120 colonies, of which half were on BS combs and the others on MD combs, situated in three different localities, were maintained over a period of five years. At the expiry of this period the overwhelming advantages of the large hive were from every point of view beyond dispute. The final change-over of all our colonies to the twelve-frame MD hives was made in 1930.

At the time these comparative tests were in progress, the view was widely propagated that in our island-climate the best results can only be secured with a bee of moderate fecundity, one that could be accommodated in a single brood chamber of BS size.

Indeed, it was asserted: 'We want honey, not bees' – clearly a case of sophistry of the worst possible kind, for we know that a large colony will, in relation to its population, collect much more honey than the same number of bees set up in two separate units. However, one important proviso is called for, there must be an appropriate balance between the nurse bees and the foraging population. On the other hand there are, of course, strains, particularly of the Italian race, that breed to excess and that will turn most of the honey they gather into brood.

At Buckfast we have at all times selected our strain with the aim of combining high fecundity with a corresponding degree of thrift. One cannot possibly hope to obtain maximum crops of honey without powerful colonies. On the other hand, it is merely asking for trouble and disappointments if a queen of exceptional fecundity is placed into a hive of restricted capacity. All too often a particular strain is blamed, when in reality it is the bee-keeper who is at fault.

The hive we have chosen serves our purpose admirably. However, I would not wish to convey the impression that I would recommend the hive we are using for general adoption. The eleven-frame MD stocked by most dealers will serve equally well. Moreover, as we have noted, good crops can be secured with almost any type of hive although not every hive will meet the requirements of easy handling and saving of time. One final remark: a hive of the size such as used by us, would merely lead to endless disappointments unless it was stocked by queens of the highest quality and strain.

4. THE APIARY

When looking for a site for an apiary we place great value on a southerly sunny aspect and shelter from the prevailing wind. An effective windbreak is most desirable, but a place that is too enclosed, with little or no circulation of air, will be avoided. We have had sites of this kind, which we had to vacate again as the bees failed to do any good. Damp sites and frost pockets are likewise avoided. Ease of approach is another important consideration. Some of our sites have been in continuous use since 1920, whilst others turned out unsatisfactory and had to be given up within a few years. To find suitable sites is no easy matter.

Experience has shown that the manner hives are positioned has a bearing on the well-being of the bees as well as on the crops obtained. For some inexplicable reason hives are commonly set out in rows with the entrances all facing in one direction. We followed this conventional arrangement at one time too, but were soon made aware of the serious drawbacks and disadvantages such a positioning of hives will inevitably entail.

When there are no definite, clear distinguishing marks between one hive and another, bees cannot readily recognise their own hive. As a result they are liable to enter, or will try to enter, the wrong hive. During a honey flow they will be admitted without getting challenged, but in periods of scarcity they will be attacked and killed if they fail to escape.

All bees are liable to drift, though there are marked differences between the various races and strains. It might perhaps be argued : what does it matter if the bees are accepted in a hive not their own ? Unfortunately, where there is any disease, drifting is the quickest and most usual way of spreading it from one hive to another. Drifting is, doubtless, likewise responsible for the loss of many queens. Finally, colonies in which drifting bees accumulate, that is those colonies at the end of a row, fail to produce crops corresponding to their strength. This is due to an imbalance and lack of harmony in the composition of such colonies. As already pointed out, where drifting takes place to any extent there can be no dependable comparative tests and no objective assessment of individual colony performance.

There are a number of means by which drifting can be reduced, but it can never be avoided completely. We favour setting up the hives in groups – four hives to a group – and the entrance of each facing in a different direction of the compass. According to tradition hives should face south-east or south, but our experience has shown that the direction a hive faces makes not the slightest difference to the amount of honey a colony will gather. There are in fact advantages in facing them in a northerly direction as such colonies will stay much more quiet in the autumn and winter.

This arrangement of setting up hives in groups possesses in addition a number of time and energy-saving advantages. The hives rest on dual stands with a space of about 8 in. between the hives. This allows us to use the adjoining hive as a table, on which the roof, crown board, feeder and super are placed whilst one hive is examined. The hive-stand, in turn, rests on a concrete platform, from which four brass pins project, corresponding to the centre of the four legs of the hive-stand. These pins fit into the legs, thus holding the hive-stand securely in place. The upper edge of the brood chamber is approximately 2 ft. off the ground, at a convenient height to facilitate the examinations with a minimum of effort. The two pairs of hives forming a group are spaced 28 in. apart.

Up to 1930 we kept as many as a hundred hives in an apiary, but now due to the more intensive farming on the one hand and the use of more powerful colonies on our part, forty seems the maximum number for best results. Rather surprisingly we have not been able to observe any difference in the yields where substantially fewer colonies are kept.

5. THE AIM OF BEE-KEEPING

In order to place bee-keeping at Buckfast in true perspective, I have to point out that South Devon is no bee paradise. Our annual rainfall averages no less than 65 in. and is one of the highest in the British Isles. The imperative need of exceptional strong colonies such as can take full advantage of the honey flow whenever it arises is a prime necessity. With haphazard methods one cannot hope to secure remunerative crops in this area. Only the most intensive form of bee-keeping will, in such environmental conditions, prove profitable.

While, therefore, our bee-keeping is of necessity carried out on intensive lines it is nevertheless based on the simplest and most elementary ways of management. Methods of questionable value and every step not really necessary are eliminated. It is indeed truly astonishing what few means we can employ that have a positive influence on the well-being and prosperity of a colony and in turn on the ultimate economic results. When all is said and done, our efforts and endeavours are restricted to a kindly ministration serving the needs of a colony. When bees were still kept in skeps the term 'bee master' was in common use – with some justification – for at that time the life and death of a colony was at the end of each season, determined by him. But we really never have had or ever will have a mastery over the honeybee. She is wild by nature and will at all times have her own way and will unfailingly and unerringly follow her instincts. It is up to us to understand her ways and adjust ourselves to her truly marvellous nature, not attempting the impossible of 'mastering' her, but rather doing all we can to serve her needs.

It will be readily appreciated that in the course of many years and daily contact with bees, the professional bee-keeper will of necessity gain a knowledge and insight into the mysterious ways of the honeybee, usually denied to the scientist in the laboratory and the amateur in possession of a few colonies. Indeed, a limited practical experience will inevitably lead to views and conclusions, which are often completely at variance to the findings of a wide practical nature. The professional bee-keeper is at all times compelled to assess things realistically and to keep an open mind in regard to every problem he may be confronted with. He is also forced to base his methods of management on concrete results and must sharply differentiate between essentials and inessentials.

One thing is certain : no person deriving his livelihood from bee-keeping will fail to take full advantage of any device or method of management which would have a bearing on an increased production per colony and an enhancement of the ultimate returns. I believe I have in the course of the past sixty years put most methods of bee-keeping to the test. At one time the most radical interventions in the organisation of a colony were deemed desirable and even necessary to success. But time and experience have unequivocally demonstrated that the bee-keeper must respect, pay a due regard to, the highly organised ways of the honeybee, if he wishes to maker her serve his best interests. But one thing seems quite clear, measures by which the professional bee-keeper ensures the best possible returns must, *ipso facto*, prove equally reliable where only a few colonies are kept.

Although some may keep bees as a hobby and relaxation, and others as a means to making a livelihood, yet all have one end in view, that of obtaining honey from their stocks. All our efforts as bee-keepers must be ordered to that goal of producing as much honey as possible. However, bee-keepers do not measure their success by bumper crops, which occur but rarely. Their true standard of success is determined by the average yield per colony over a period of years.

This is the Rule of the Golden Mean. But, the aim of a high average yield per colony raises a whole series of questions: first and foremost the development of a type of bee possessing the ability of realising such an objective. We are thus at the outset up against the all important question of bee breeding, the heart and central issue of our bee-keeping. All that we do in our apiaries throughout the year is subordinate to this all important problem: the creation of a bee endowed with the qualifications that will enable it to fulfil the aim we have set ourselves. To assist that bee to attain our objective we have in turn to provide a seasonal management adapted to her particular needs.

PART II
SEASONAL MANAGEMENT

The previous chapter was devoted to an account of some of the basic principles which form the foundation of bee-keeping at Buckfast. These basic principles have been ascertained and verified by a practical experience of bee-keeping extending over a period of well-nigh sixty years. This experience has been gained in the stern school of practical everyday bee-keeping and in an unusually difficult climatic environment. The exceptionally high rainfall, averaging 65 in. in our particular part of Devon, is inevitably reflected in the problems and uncertainties entailed in the production of honey. The high hopes in spring are more often dashed by a wretched summer; the laborious building up of colonies during the summer is all too often in vain when the heather on the Moor fails to yield because of long spells of dull, drizzly and misty conditions. In spite of this the work must begin in the spring and proceed systematically until the autumn. Optimism, patience, perseverance and hope are some of the qualities a bee-keeper must have.

In this chapter the various stages of our method of managing the bees through the different seasons will be described.

1. SPRING

a. First inspection of colonies

From the beginning of October until the end of February the bees are left to their fate. The whole of this period will often pass by without a visit to any of the apiaries. In the autumn every colony is provided with a sufficient quantity of stores, mostly honey, to ensure its survival until mid-April. The hives and roofs are firmly secured in position by a strand of wire fastened to the hivestand, so the fiercest of winter storms causes us no concern. Towards the end of February, or as soon as conditions thereafter permit, steps are taken to clean the floor boards and to ascertain a cursory estimate of the condition of each colony.

I have already mentioned that forty colonies are kept at the majority of our apiaries, so an equal number of bottom boards is held in reserve. Each morning, weather permitting, an apiary is visited and the bottom boards changed for a clean set. The boards with the debris and bees that died during the winter months are brought home and scrubbed in boiling water, containing some detergent, then dried overnight. By the middle of March this task is usually completed. By that time it is generally warm enough to permit a closer check of each colony, and we take this opportunity to remove any combs not covered by the bees. A note is made of the number of combs each colony covers and from these notes the overall average strength is assessed. Thus, we know in advance which colonies are in need of help, the exact amount of help they require, and at the same time which of the colonies can give up combs of bees and brood.

b. Equalising

Equalising means attempting to establish all colonies throughout the apiary or apiaries at the same level of strength, so that at a given date in early spring all the colonies will be starting the season on a footing of equality. Obviously judging the strength of a colony is something which requires an experienced eye, but it is an art that can be readily acquired. However, it is true to say that the ability to equalise colonies effectively may be regarded as the hall-mark of an accomplished bee-keeper.

The reason for our insistence on the importance of equalising is not some exaggerated sense of neatness and tidiness, but the bearing it has on the overall development of the colonies and their performance. Apart from the economic advantages gained there are, in addition, many of a practical nature. When all the colonies in an apiary are of uniform strength, the apiary can be dealt with as a unit. If one colony needs more room, all the others are likely to do so. This uniformity simplifies management immensely. Where many apiaries and a large number of colonies need attention, there is nothing more time-consuming than the care of colonies that are either below or above average strength. The former, left to themselves, will mostly fail to reach full strength in time for the main honey flow ; the latter, on the other hand, will usually dissipate their above average strength in swarming and then fail to make the most of a honey flow. As already indicated, theoretical considerations confirmed by practical experience have clearly shown that the overall strength will be substantially greater in time for the main honey flow than if the equalisation had not been made.

Equalising as here outlined can, of course, only be carried out successfully when there is more than one apiary at the disposal of the bee-keeper. The bees can then be taken to a different apiary, preventing them from returning to their parent colony. Carried out thus the results of the equalisation are in no doubt and can in fact be determined with certainty. But where the bees are kept only in one place the equalising has to be postponed until an abundance of young bees are present, as these will not return to their original home. However, in such instances the equalising will have to be performed with more care and circumspection to ensure success, for the number of bees that will return to their parent colony cannot be determined with any certainty in advance.

Before any combs of brood with the adhering bees can be taken away from the colonies above average strength the queen must of course be found in every case. We therefore do our regular annual requeening in conjunction with the equalising. The majority of queens have to be found inevitably, so both these tasks are done at one and the same time thereby ensuring a substantial saving in labour.

c. Requeening

Although we requeen colonies whenever deemed desirable at any time of the season (this will be discussed at length in due course), it is in the spring that we do the main requeening for a number of reasons, apart from those already mentioned. First, a few words on the method we have adopted.

We usually winter a large reserve of young queens at our

mating station situated in the heart of Dartmoor. The queens are wintered in nuclei on four half-size MD combs, equalling the comb area of three BS frames. They have consequently to pass a severe preliminary test of endurance in the harsh wintry conditions of Dartmoor, at 1,200 ft. above sea-level, before they are transferred to the honey producing colonies.

Each morning, when the requeening and equalisation is done, we first visit the mating station armed with the number of queen cages needed for the day's requeening. These are of our own design and construction and measure 3¾ in. x 1 in. x ½ in. and are made of 12 x 12 x 30 mesh gauge wire gauze closed at both ends with a block of wood ½ in. in width. One of the blocks has a ⅜ in. diameter hole, by which the queen and three or four attendant bees are inserted and then confined with a plug of candy. After the required number of queens have been caged we set out for the apiaries to be requeened. The old queens in the honey producing colonies are caged in the same manner and the young queens introduced simultaneously by wedging the wire cage between the top bars in the centre of the brood-nest. The bees will consume the candy and liberate the young queen in the space of a few hours whereupon she will resume her egg-laying as if nothing had happened. When the day's requeening and equalisation is finished, a second trip is made to the mating station when the old queens are given to the nuclei from which the young queens were taken away in the morning. These old queens will remain in the nuclei until the end of May when they will be removed to make way for the virgin queens raised in accordance with the year's breeding plans. This procedure is followed every day until the honey producing colonies in need of new queens have been requeened and the equalising is completed. Approximately two-thirds of our colonies are requeened annually.

The procedure I have outlined inevitably poses a number of questions: first and foremost, why is the general requeening done in spring and not in July when the season's young queens are available?; secondly, how is it that so many colonies can be requeened without any fear of valuable queens being lost? These and all relevant questions are perhaps best answered by a consideration of the principles which lie behind successful queen introduction.

31

d. Queen introduction

The queen is obviously the primary source of the well-being and productivity of a colony. By the mere substitution of a young queen we are enabled to renew the main-spring of the life of a colony – indeed, we have at our command the power to rejuvenate and keep a colony perpetually young and at its maximum productive ability. A dependable method of getting queens accepted is therefore essential wherever high average crops per colony are under consideration. Indeed, I regard this question of the introduction of queens as one of the few things that matter and a pivotal point in modern bee-keeping.

My views and findings in regard to the introduction of queens are fairly well known. They were first set down in writing more than thirty years ago and published in full in 1950. A summary was submitted to the International Bee-keepers Congress held in Leamington Spa. In the intervening years no evidence has come to light which would invalidate my findings. We have meanwhile also been able to check the re-action of almost all known races and crosses of bees of which I had no previous experience.

Put briefly, my contention is that the acceptance of a queen is not determined, as hitherto generally assumed, by 'colony odour', but by her behaviour. A fully mature queen, one that has been laying for a considerable time, will have lost her original nervousness and will behave sedately and calmly. When in that condition, her acceptance is assured irrespective of the safeguards generally considered as essential. Odour, or 'colony odour' – if there is such a thing, which I doubt – plays no role in the acceptance of a queen. The essential condition which ensures acceptance or rejection is in the final analysis determined by the behaviour of the queen. The behaviour of a queen is in turn dependent on her condition at the time of liberation.

It is not possible to discuss here at length the very interesting theoretical considerations in the hypothesis I have put forward. I am here primarily concerned with the stricly practical considerations and practical application of my supposition. To ensure a better understanding of the problems at issue I have, however, to point out that the two terms 'colony odour' and 'hive odour' are often assumed to be one and the same thing.

While a number of eminent scientists have at one time or another claimed they had successfully traced the source of 'colony odour', these discoveries have unfortunately not found any corroboration in the field of practical experience. We have in fact no positive evidence that there is a 'colony odour'. If there should be a 'colony odour' it would of necessity be based on hereditary factors. Keeping in mind the known reactions on the part of individual colonies, a moment's reflection will show that this could not possibly be the case. One thing, however, seems certain: 'colony odour' – if there should be such a thing – has nothing to do with the rejection or acceptance of a queen.

On the other hand, 'hive odour' based on a combination of odours emanating from the combs, brood, honey, pollen, propolis, etc. is an obvious reality. But the difference between one hive and another, in the same place and identical environment, can hardly have any bearing on the acceptance of queens. Indeed, we know it has none whatever.

Backed by a lifetime's experience I am firmly convinced that, whatever the method of introduction employed, the factor which determines acceptance or rejection in introduction is primarily the behaviour of the queen in every case. The behaviour of the queen is in turn subject to the condition she is in at the time when she is liberated.

A virgin or newly-mated queen is as a rule extremely nervous and easily frightened. The slightest disturbance, the mere opening of the hive, may place her life in jeopardy. In the course of a few weeks, after she has commenced to lay, a radical change in her behaviour will manifest itself. Her movements will be more deliberate and sedate. By the time she is surrounded by her own offspring, four or five weeks after she commenced to lay, she will have reached her first stage of maturity. Her maximum laying capacity will, however, not be attained until the following year. But in her behaviour no further marked difference will be manifested. The time indicated for the attainment of the first stage of maturity, viz. four to six weeks, will in cases of an innate nervousness and with certain hybrids have to be extended. According to my experience, two months seems to suffice in every case. The western European black bee, especially the southern French and Iberian sub-varieties, appear the most intractable in this respect.

I have used the conventional term 'introduction' but in reality the procedure merely involves an 'exchange' or 'substitution' of one queen for another. There is no preliminary 'getting used to' or an 'acquiring of colony odour' entailed before a queen is liberated. A substituted queen will immediately on being freed resume her normal activities regardless of her new surroundings – just as a bee returning from the fields laden with nectar or pollen, on missing her own hive, carries on in her new surroundings as if she had entered her rightful home. A substituted queen is accepted as the rightful mother of a colony solely by virtue of her condition and behaviour.

It should be clearly understood that queens which are not in laying condition, such as have come through the post, cannot be thus given to a colony. We introduce such queens first to small nuclei, composed mainly of young bees, and keep them confined for twenty-four hours before liberation. When again in laying condition they can be transferred to large colonies.

To my mind there is no single more unfortunate factor in modern bee-keeping than the very high loss of valuable queens universally sustained by the methods of introduction commonly advocated. There is, in addition, an aspect of introduction which must not be lost sight of, namely, all too often queens that are accepted suffer injuries to an extent that they will be superseded within a few weeks or months without the bee-keeper being aware of it – unless they have been marked in some way or another. Many queens are likewise not visibly injured, but have been harmed nevertheless, with the result that such colonies fail to prosper and fail to reach the expected productivity. Indeed, a colony headed by a defective queen is generally of no real practical value. In fact, queens of this kind are more often than not a source of endless trouble.

e. The time for requeening

We may now return to consider some of the questions about requeening which were raised earlier and which can now be more readily answered in the light of what has been said about the introduction of queens.

We maintained that a queen will never do as well in the season she is born as when she has reached her prime in the following year. So if at all possible we never make use of young queens in the season they are born. The view is commonly held that young queens are essential to ensure safe wintering of colonies, and also that they are indispensable for building up strong colonies for obtaining the best results on the Moor. These views doubtless hold good where a colony is headed by an old or failing queen, but all our experience has shown that the most populous colonies with the strongest force of field bees for gathering the heather crop are provided by queens in their prime, that is in the year following their birth. The same holds good in regard to wintering. Of course, if at any time a colony is queenless, or in possession of a failing one, and our reserve of queens of the previous year is exhausted, we make use of newly-mated queens.

The advantages of requeening in early spring are so many as to neutralise the objections that can be raised against the scheme I have outlined. Let us consider some of the main advantages:

(i) requeening is done in conjunction with the equalisation and at a time of the year when no other important work demands our attention;

(ii) as queens at this time of the year are most easily found, the labour entailed is reduced to a minimum;

(iii) the change of queens is effected with practically no disturbance to or interference with the colonies;

(iv) there are no losses and all the queens are accepted without their sustaining any injury.

We have carried out our annual requeening on these lines since 1930. The wintering of a large reserve of young queens may at first sight seem an extravagant use of equipment, bees and honey, and an unwarranted upkeep and outlay. This is, however, not the case, for the nuclei are almost self-supporting and permit us to subject these queens to a preliminary test before they are transferred to the honey-producing colonies in the spring. This allows us to eliminate at the outset any queen, or possibly a line of queens, which have not proved completely satisfactory. But first and foremost this reserve of queens enables us to do our re-queening at a period of the season involving a minimum of effort and time, and when acceptance is virtually infallible and the possibility of injury to the queens nil.

All that has been said so far presupposes the possession of a number of colonies if not several apiaries, and this may seem far removed from the situation in which the ordinary amateur beekeeper finds himself in this country. But again let it be said that in this book I am trying to bring out the principles of bee-keeping by describing our methods at Buckfast. These principles remain valid even though their application to individual practical needs will vary.

f. Uniting

I cannot conclude this section without a word on uniting, or more precisely on the transfer of bees from one colony to another when the equalisation is performed. The danger of the bees fighting and the role played by 'colony odour' will have doubtless come to the mind of many a reader. Here again 'colony odour' (if it exists) is of no significance in the uniting of bees of different colonies. The factor making for success is once more the behaviour of the bees. Every bee-keeper is aware that exposure to light has a calming effect on bees, and any which have thus been exposed for some minutes will peaceably join with bees of other colonies without the need for any other precaution. During the whole of the season, when for any reason a transference of bees from one colony to another is necessary, we employ no other safeguard to prevent fighting than exposure to light. When dealing with mongrels or races of exceptional nervous disposition extra care is called for.

A summary of the main points made in this section may prove helpful :—

(i) The assumption that 'colony odour' has any significance in queen introduction is in our view without foundation in fact.

(ii) The determining factor in acceptance of a queen is her behaviour at the time of her release. If her behaviour is such as not to arouse the hostility of the bees, she will be accepted without fail.

(iii) This behaviour of the queen is dependent on her condition and age. If she is mature and in full laying condition when released, she will immediately resume laying as if nothing has transpired.

This section of queen introduction has been a lengthy one, but I am here dealing with a matter of the utmost importance. Every year a high percentage of young queens perish on the threshold of their useful existence through mistakes in introduction. A method of getting queens unfailingly accepted, and accepted without their suffering any injury, is an essential prerequisite of successful beekeeping. In the section of this book dealing with queen rearing special emphasis is placed on the overriding need of queens of the highest quality, but this would be labour lost if there were not also available a completely reliable method of introducing them to the colonies they are to head.

2. SUMMER

a. The build-up

By the time the annual requeening has been completed about two-thirds of our colonies are headed by young queens; that is, the queens raised the previous summer and wintered at the mating station in the heart of Dartmoor. Those queens which have proved themselves as the best in the previous season remain in the honey-producing colonies a further year. The most outstanding are reserved as breeders.

When the equalising is finished all are generally on seven MD combs. But this number varies from year to year, as the average colony strength in early spring is largely determined by the nature of the honey flow on Dartmoor the previous autumn. Subsequent to a total failure of the heather – an event that will happen now and again – colonies may be in such a weak condition as to endanger their survival. In 1946 we had such a season with the result that in the spring of 1947 our colonies barely covered four combs after the equalisation was completed. On the other hand, exceptional colony strength is not desirable at this time of the year, for we have no spring honey flow of any importance. Colonies of medium strength will usually manifest the most satisfactory build-up and will reach the peak of their strength at the right time to permit them to take full advantage of the main honey flow from the clover. Indeed, a strong nucleus of the previous year will often build up more rapidly and collect more honey than colonies of abnormal strength in the early spring.

After the equalisation and requeening, colonies are left undisturbed until the middle of April. During this period of the season they are kept restricted to the number of combs they can fully cover. When thus managed the development will go forward more rapidly. The risk of an outbreak of Nosema will likewise be thus considerably minimised, if not completely avoided.

Unless the weather has been particularly unfavourable, the colonies will by mid-April be in need of an additional comb and a further one about ten days subsequently. This giving of additional space is done step-by-step and by the end of May or early in June each colony should be in possession of its full complement of twelve combs. All the combs are added on the periphery of the broodnest, alternately next to the division board and the far side of the brood-nest. The expansion of the brood-nest must be spontaneous and without the application of any compulsion on our part. The spreading of brood by the insertion of empty combs or frames fitted with foundations in the middle of the brood-nest, as widely advocated at one time, violates evey convention of good beekeeping.

The placing of additional combs or foundations on the periphery of the brood-nest possesses many practical advantages. Given this way they can be inserted at any time convenient to the beekeeper, without incurring any risk of damage to the foundations. When added on the periphery the bees can take possession of them and draw them out when they feel so disposed. Indeed, they will generally draw them out to perfection at this point – anyway, bees of pure stock. In the case of hybrids we are usually faced by a totally different problem. These tend to construct far too great a proportion of drone comb when the foundation is situated on the fringe of the brood-nest. But when placed between combs of brood, they will draw them out free of drone comb. As we place great value on perfect combs we have been forced to accept a compromise solution. The foundations are inserted in the middle of the brood-nest and transferred to the periphery as soon as they are drawn out. If left in the middle, they tend to form a barrier to the queen with the result that she will be inclined to restrict laying to the combs on one side of the foundation until a good honey flow supervenes. It may also occasionally give rise to swarming. But by transferring the newly drawn foundations to the periphery, these drawbacks are circumvented at the cost of considerable extra work. It is admittedly a compromise solution, but one we have to accept in the circumstances. Needless to say, should there be no honey flow when the foundations are inserted, the colonies have to be fed in order to get them drawn out. All our colonies have to draw out a minimum of three foundations in the brood chamber annually.

b. The problem of feeding

The reader will by now have drawn the conclusion that stimulative feeding, apart from getting the foundations drawn out in the brood chamber, plays no part in our scheme of bee-keeping. This is in fact so. There was a time however, admittedly many years ago, when we too considered stimulative feeding in spring desirable if not essential. In compliance with convention the feeding of candy also occupied an important role in our seasonal management at one time. From March onwards small daily doses of warm syrup in the evenings took the place of candy. No feeding is now done except when absolutely necessary and, of course, none whatever during the winter months.

We find that feeding at any time in the early part of the season will cause the raising of an undue amount of drone brood. Later on it is liable to give rise to swarming. Up to 1936 we could always rely on a favourable May and June, but the reverse has been the case with few exceptions ever since. However, seasons of complete failure have arisen at all times, possibly more often here in the south-west than in other parts of the British Isles. When feeding cannot be avoided, a substantial dose is then given – usually a gallon at a time. This way a considerably saving of time is achieved. Moreover, as our experience has again and again demonstrated, a heavy feed will prove far more satisfactory than the feeding of the same amount of syrup in small quantities.

c. Our method of feeding

The question of the method of feeding bees is an important one if the bee-keeper wishes to save himself much labour and time. We have indeed give the economic and practical aspects of feeding a great deal of consideration and in the course of time eliminated every non-essential step and have reduced this task to a point where it can hardly be made any simpler.

Before 1917 we relied entirely on candy, the method of feeding then in common use. This form of feeding is doubtless the most expensive imaginable. For this reason we adopted in 1917 the American example and started using feeder tins. We made use of the standard 28 lb. lever-lid honey tins. The lids were punctured with a series of small holes by which, when inverted over the feed-hole of the crown board, the bees gained access to the syrup. This was a great step forward on any of the methods of feeding in use at the time. These tins had, however, a number of drawbacks. For one thing, the method entailed using two empty supers as a protection for the tins. With a large number of colonies this was for obvious practical reasons an impossibility. Again, in the early morning when the temperature rose quickly, the vacuum was temporarily lost with the result the bees were swamped with the escaping syrup. This always inevitably happened when the level of the syrup was down to a quart or two. Moreover, the smallest hole in the tin caused by rust meant that there was no vacuum and the entire contents escaped over the bees and combs after the tin was inverted over the feed-hole.

The climax came in the autumn of 1932. A visit to the out-apiaries after a stormy night presented us with a scene of near-chaos. Tins and roofs had been blown all over the ground, and it was impossible to tell how much of the syrup each colony had taken down. The idea then came of using a feeder in the form of a tray which would fit on top of the brood chamber, possessing the same external dimensions and a capacity of about 1½ gallons. This tray-feeder is made entirely of wood, apart from the tin cover which fits over the central feeder block, and rendered water-proof by immersion in a tank of hot paraffin wax. The crown board and roof fit over the feeder, keeping everything secure. In the first feeders of this type an arrangement was incorporated for stimulative and rapid feeding, but the slow feeding device was rarely used and has now been abandoned. When the tray is almost empty the bees have access to the whole of the tray, enabling them to clean up the last trace of syrup. This type of feeder has proved pre-eminently satisfactory at our mating station for feeding the nuclei. The trays in this case are fitted with two feeder blocks, suitably modified to permit four nuclei to take down the syrup from the one tray simultaneously. When empty, one of the nuclei has access to the whole of the tray for cleaning it. The introduction of this feeder solved in the simplest possible manner a whole series of problems which up to then seemed insoluble.

Up to the time of the introduction of the tray-feeder we prepared the syrup as then prescribed by tradition: it was boiled and fed when lukewarm. This again entailed a process involving much time and needless labour and the use of fuel to little or no purpose. The feeding of warm syrup had in addition a number of serious drawbacks; whereas the feeding of cold syrup possessed many practical advantages, apart from the saving of time, labour and the cost of fuel. Warm syrup has to be fed in the evening to avoid excessive excitement and the risk of robbing; cold syrup causes little or no excitement and can therefore be given at any time of the day with impunity, an invaluable advantage where the feeding of a great number of colonies has to be considered.

The preparation of the syrup could again not possibly be more simple — it is in fact merely a question of dissolving the granulated sugar in a corresponding quantity of cold water. We make it in a concentration of 1 pint of water to 2 lb. of sugar. We use for this purpose a specially built open tank, lined with glazed tiles, holding about 1½ tons of syrup. The correct quantity of water is first allowed to flow into the tank and then a corresponding number of 1 cwt. bags of sugar are tipped into the water. As the sugar is poured in the solution is kept agitated by means of a 16 in. wide steel scraper adapted for this purpose for approximately 15 minutes. The important thing is not to allow the sugar to settle to the bottom, from where it will be difficult to raise as it tends to form a dense paste. At the expiry of 15 minutes, the syrup will still be cloudy but is nevertheless ready for use. If left to stand for an hour it will turn perfectly clear, exactly as if it had been boiled. The syrup is pumped from the mixing tank into a 60-gallon cask for transportation to the apiaries. To prevent any movement of the cask whilst in transit a special cradle is provided, which holds the cask firmly in position. On arrival at the apiaries the syrup is drawn off into 4-gallon cans from which the feeders are filled.

The feeding of sugar seems an inescapable necessity in the climatic conditions prevailing in the British Isles and most parts of Europe. It is, however, a measure we take recourse to only when absolutely compelled to do so. A highly efficient installation for preparing the syrup and feeding are imperative where time and labour form an important consideration. Our method of preparing syrup and the use of the tray-feeder enables us to feed 320 colonies, dispersed in nine apiaries, within a matter of eight hours.

A timely feed of syrup will often mean a difference between success and failure or even disaster. But there is also little doubt that in the hands of many well-meaning amateurs the feeding of sugar is liable to be abused.

d. Supering and renewal of combs

About 20th May our colonies are usually ready for their first super. This is given before they have the full complement of twelve combs in the brood chamber. We find if the supering is delayed until they have the whole set of combs, signs of swarming will be manifested before the end of the month. On the other hand, if the first super is given as soon as the colonies fully cover nine combs, the build-up will go forward without any interruption.

For close on twenty years we did not use any queen excluders. We found, however, that the drawbacks were very much greater than any advantages their absence may have had. Wire excluders of strong construction, the type we now use, will not materially hinder or interfere in the bees entering the supers as was the case when the old-time zinc excluders were used.

As already indicated, all our colonies have to draw out a minimum of three foundations in the brood chamber annually. They have in addition to draw out a great many in the supers, indeed now and again the entire set of supers subsequent to a good crop of heather honey.

The periodic renewal of brood combs is doubtless one of the most effective disease-preventive measures – a measure widely neglected in modern bee-keeping. Apart from Acarine the source of most diseases is carried in the combs, primarily in the brood combs. We have, in addition, to take the progressive spontaneous deterioration of the brood combs into account. A complete renewal, in one operation, every four years, would in my estimation prove the ideal solution; that is, as a disease-preventive measure. We did in fact at one time adopt such a procedure, completely renewing in one operation the whole set of brood combs of eighty colonies in the four-yearly rotations.

We are apt to forget that, up to the time of the introduction of the movable-comb hive, the periodical renewal of combs took place as a matter of course. Every autumn the heaviest colonies were set over the sulphur-pit and the combs subsequently destroyed. We may safely assume that bee diseases were at that time far less common than at present. The failure to replace combs periodically – notwithstanding the imperative need on hygiene as well as practical grounds – seems one of the most vulnerable aspects of modern bee-keeping. While at certain periods of the season queens and bees manifest a preference for old combs, colonies on a new set will usually evince a zest and prosperity lacking in those on old combs.

The procedure adopted by us for the total renewal of combs was as follows: as soon as the colonies reached their peak of strength, viz. towards the end of June, just prior to the main honey flow, one-quarter of our colonies were transferred on to a set of frames fitted with foundations and into hives which had been previously sterilised. If there happened to be no flow at the time, these colonies had then to be fed in order to get the foundation drawn out quickly. Enough bees were left with the brood, which was set above the crown board and the brood chamber with the foundations, but provided with a separate entrance. Ten days later the majority of the young bees which had meanwhile emerged were returned to the parent colony and all the queen cells on the old combs destroyed at the same time. Twelve days later, by which time all the brood will have emerged, the remaining bees were reunited to the parent colony and the empty combs stored away for melting down.

We found that the changeover as just described did not adversely affect the crop from the clover – indeed, rather the opposite – but it severely curtailed the one from the heather, and to such an extent as to compel us to look around for an alternative. The cause of the failure on the heather was due to the break in brood rearing following the changeover. While the actual cessation did not last for more than a day or two, the amount of brood raised during the critical period, when the bees that gather the crop on the moor are reared, could not possibly equal that of a colony which suffered no disturbance of any kind. In other words, the bees needed to gather a maximum crop from the heather were not available.

A compromise solution had to be found, as is so often the case in bee-keeping. For many years the following alternative has served our requirements. Three of the oldest combs are replaced annually; these are transferred at the end of June to the fringe of the brood nest, one to the near and the other two on to the far side of the brood chamber. In the following spring, or as soon as convenient, they are removed and melted down. The disorganisation caused by the rearranging of the combs, at the end of June, will occasionally give rise to swarming. But as this seems the only feasible way we can accomplish a periodical replacement of brood combs, which we consider necessary, we are left with no other choice but to accept this minor drawback. The general health and prosperity of the colonies is of far greater importance; and we have at all times to consider the overall advantages.

It might well be argued that the periodic renewal of combs and the cost of the conversion of the wax into foundation would prove an expensive undertaking. However, we have found that, with a moderately efficient wax rendering plant, the greater quantity of wax secured will defray the expenses incurred.

e. Periodical inspections and swarm control

Our way of bee-keeping, necessitated by the particular environmental conditions, calls for a close control of colonies from the end of March until the conclusion of the swarming season. We rarely allow more than a fortnight to elapse from one check to another. Thus, we are able to keep our fingers on the pulse on the condition of each colony. These frequent inspections enable us above all to assess the value of each queen. If one does not come up to standard, she is replaced at once, irrespective of her age.

However, it does happen that due to no fault on her part a colony will not make the progress expected immediately after the requeening in March until her own progeny have taken over. Such cases are not always easy to determine and a number of factors have to be taken into consideration. As a help, the actual date of the introduction is noted in each instance. These periodic checks take up very little time, except where swarming preparations are actually under way. In all other cases a quick glance at two or three combs will tell us whether everything is in order.

At the time the old native bee was still extant the first swarms usually issued before the end of April. With the present-day bee, and more advanced methods of bee-keeping, swarming is, in our apiaries, virtually unknown before the end of June. Indeed, the main swarming season is confined to the period between 5th July and 25th July. Furthermore, only a small number of colonies will normally manifest any swarming preparations, and a still fewer number will actually attempt to swarm. But even so, swarming must be regarded as one of the foremost obstacles to successful bee-keeping.

There are, admittedly, an endless number of swarm preventive measures; none can be relied on with any certainty, apart from the one involving the removal of the queen for a period of ten to fourteen days. This was in actual fact the method we employed at one time. It is a method that possesses many advantages of both economic and practical importance besides providing the only positive control over swarming. It is also perhaps not generally realised that a spell of queenlessness, just prior to the main honey flow, will help to check both adult and brood diseases. Indeed, it is one of nature's ways of controlling bee diseases.

We applied this method to all colonies, irrespective of whether swarm preparations were in progress or not, on or about 21st June. Ten days later all the queen cells were removed and a few days after, a young fertile queen introduced. By the use of this method the weekly examinations were avoided. Furthermore, a substantial increase in the crop from the clover was thus secured, for as soon as unsealed brood was again present, colonies so treated worked with an energy and determination normally manifested only by newly-hived swarms.

46

As already indicated, this swarm-preventive measure posses-ses a number of additional advantages. The break in brood-rearing has a prophylactic influence on colony health. It also brings about a reduction in colony strength subsequent to the main honey flow at a time when an excess of strength can be a positive drawback; that is, in areas without a late honey flow. Colonies thus treated will go into winter in a rejuvenated condition and, as experience has demonstrated, will forge ahead in the spring build-up with a vigour missing in colonies not so managed.

This method of swarm control excludes any uncertainty. There is no question: will it, or will it not work? The many other advantages are likewise equally certain. Unfortunately, as we soon found to our loss, on Dartmoor one cannot hope to obtain a worthwhile crop except from colonies at their peak of strength. Here again the break in brood rearing, at a time when the bees should have been raised for gathering the crop in the latter part of August, proved responsible for the loss of the necessary colony strength. We were therefore left with no other option but to resort to a method of swarm control which ensured an optimum strength of the majority of the colonies.

In the circumstances only one course was left open to us: a weekly check of all colonies during the swarming period and removal of all queen cells where any are found. This is, admit-tedly, not an ideal solution, but one which has given acceptable results over many years. We find these weekly checks take about four minutes on an average per colony. Where any queen cells are found, they are destroyed. If the queen still lays normally, there is every likelihood that in the following week all preparations for swarming will have come to an end. On the other hand, when a queen has stopped laying, such colonies will usually make one or two attempts at swarming. But as a wing of the queen is clipped, the swarm will be compelled to return. With the type of alighting boards we have, which reach down to the ground, the queen will in most cases crawl back into the hive, but some are inevitably lost. Colonies that have lost their queens are requeened a week or ten days later. There would be no point in giving them another queen immediately, for they would most likely attempt to swarm with the new queen. A period of queenlessness is essential, prior to the introduction of another queen. Such colonies will, of course, prove of little value on the heather.

I should have pointed out, before now, that we never introduce a queen without first clipping her wings. Needless to say, ·this will not prevent attempts at swarming, but a swarm cannot decamp without a queen. The clipping of wings will not only prevent the loss of swarms but will also avoid the loss of time and the risks entailed in collecting them, particularly where tall trees are situated in close proximity to an apiary. The fears now and again expressed in regard to the possible injury a queen may suffer by the clipping of her wings, resulting in an early supersedure, have not been substantiated by experience. I have in no case been able to observe any harmful results, traceable to the clipping, in sixty years of bee-keeping. However, if awkwardly handled and clipped a queen might suffer permanent injury. The successful keeping of many colonies in a series of out-apiaries would obviously prove virtually impossible without this safeguard – except perhaps in the case of a strain of an unusual non-swarming disposition. We regard the clipping of wings – only one-half of one wing is removed – as an elementary commonsense precautionary measure. But there is no point in clipping, except in conjunction with a weekly check of colonies during the swarming season. If this is not done, a swarm will issue and depart with a virgin queen.

It must be remembered the swarm-preventive method used by us is based on a strain in which the non-swarming disposition has been highly developed. A method of this kind would prove of little or no practical value in the case of the Carniolan bee, certain hybrids and the generality of mongrels. In instances such as this only extreme measures, of a radical nature, will serve any purpose.

f. Routine procedures during the honey flow

The weekly inspections during the swarming season coincide with the main honey flow. They also provide an opportunity for giving additional supers where found necessary. We rarely have a reserve of drawn combs and the supers are mostly – in some years entirely – fitted with foundations. They are consequently always added on top of those already in position, for here the foundation is drawn out more expeditiously, often within a few hours when there is a good honey flow. By adding them on top much lifting is obviated and little or no disturbance caused to the colonies. Furthermore, if the weather should unexpectedly deteriorate, no harm will result. On the other hand, during a heavy honey flow an additional check on super room may be called for, in which case a quick glance by merely lifting the crown board will show in an instant whether more storage space is needed. Towards the end of the honey flow and at the time the last weekly check is made, the partially filled super is placed next to the queen excluder and the full ones on top. This allows the removal of the full supers without further disturbance. Also, towards the end of the flow, the bees will tend to store the incoming nectar close to the brood-nest.

For clearing the bees from the supers we use Porter escapes. This device causes less of an upset to the colonies than the chemical or mechanical alternatives introduced in recent years. When the combs are fully sealed, bees will often leave the supers within a few hours in warm weather. However, they are usually left two days on the escapes, to ensure all are completely free of bees.

The extraction and removal of the supers is generally put off until the end of the honey flow, or until empty supers are required for preparing colonies for the heather. At the conclusion of a good season the difference in individual colony yields will be visible at first sight. But with supers holding 50 lb. of honey each, stacks reminiscent of American bee-keeping are absent. The maximum heights attained are in every case confined within easy reach — an advantage of no small importance.

g. Heather honey production

The fact that Buckfast is within easy reach of Dartmoor gives us an opportunity of obtaining a crop of honey from the heather. But the best advantage can be taken from this only by careful planning, so much so that all our bee-keeping during the summer has the heather honey harvest constantly in mind. Only the strongest colonies and most vigorous of bees can be expected to produce good results in the conditions which prevail on the Moor, where even in August and September the bees can be exposed to very rough conditions and adequate shelter is hard to obtain.

I can recall a time when we transported colonies to the fringe of Dartmoor on a wheelbarrow. In the course of the years the mode of transportation was progressively rendered less arduous and far more expeditious. The hardships, setbacks and disasters of these early endeavours now form but distant memories.

Success in heather honey production will only come to the determined and persevering bee-keeper, for very much more is entailed than merely placing the hives within reach of the heather. In the course of the past fifty-five years I observed numerous attempts, based on a supposition of this kind, resulting in endless disappointments. Indeed, I cannot recollect a single bee-keeper who persisted over many years. Taking bees to the heather entails heavy work, many risks and many failures. In seasons of total failure, the near ruination of the colonies may have to be accepted. The loss of colonies will not come immediately, but in the course of the following spring. Where much heather honey is stored in the brood chamber, widespread cases of dysentery are an ever present danger. These dangers and risks certainly hold good for Dartmoor, but possibly to a lesser degree to moorland areas in other parts of the British Isles.

Success on Dartmoor demands a combination of conditions, which many bee-keepers are unable to provide or fail to recognise. I will cite a few classical examples in due course. The concept of 'success' is admittedly a very relative one, as success in the eyes of one person may be deemed a failure by another.

Most years the first clumps of ling will be in bloom on or about 25th July. A further three weeks will usually pass by before the bulk is in flower. Only in very exceptional seasons have I known it to secrete freely before mid-August and after 5th September. The flow often sets in suddenly and as a rule will last but a few days. Close, sultry, windless days and warm nights seem most conducive to a heavy flow. The ling will secrete at relatively low temperatures but will cease abruptly with a change of wind

50

to a north-easterly direction. On the other hand, when conditions are just right, the secretion can only be described as stupendous and amazing.

We generally start with the transportation of the hives to the heather on or about 28th July and as one apiary is moved each morning the operation lasts about ten days. The loading and transportation is done at daybreak and all necessary preparations for fastening the hives, the previous afternoon.

For more than fifty years we have used a very simple and exceedingly reliable method of securing the hives for transportation. On the underside of each bottom board are attached permanently two brass plates, one at the back and the other at the front. They measure 2 in. x 1 in. x ⅜ in., and are fitted with a ¼ in. threaded hole in the centre. In the wooden frame of the screen are likewise two ⅛ in. thick steel plates, each with a ⅜ in. hole in corresponding positions to the threaded part of the brass plates fixed to the bottom boards. Two ¼ in. diameter steel rods, threaded one end and the other fitted with a wing nut, comprise the fastening device. The two steel rods are inserted into the plates attached to the screens and allowed to drop along the inner wall of the hive and screwed into the brass plates of the bottom boards. When screwed home the bottom board, brood chamber, queen excluder, super and screen are firmly clamped together. The rod next to the entrance is not immediately fully screwed home. On arrival next morning a slip of wood is inserted into the entrance and by screwing the rod home fully the entrance block and closing slip are firmly held in place. The hives are then ready for loading.

When transported to the heather each hive is provided with one super; before they are brought back, the supers are removed and taken home separately. Two sets of steel rods are therefore required: one set long enough to include a super and a short set for the return trip. I should perhaps point out that the wing nuts are fixed to the steel rods and do not revolve; and both sets should be about ¼ in. longer than actually needed, to allow for a possible expansion in the depth of the brood chamber and super in rainy weather. This arrangement for securing hives for transportation is simple, absolutely dependable and cannot cause any damage to the hives.

It will be appreciated, a completely reliable means of fastening the hive components, and the elimination of any possible risk of bees escaping en route, is an elementary necessity when bees are transported by road. Unreliable makeshift devices will sooner or later lead to unpleasant experiences.

Management on the Moor is restricted to the giving of additional super room when called for or warranted by the honey flow and the weather. Owing to the many imponderable factors at issue the giving of more room demands good judgement and an equal measure of luck. We do not wish to be left with a great many unsealed and half-filled combs at the conclusion of the season, which apart from the loss of combs, detracts from the quality of the honey. The additional supers are in every case placed on top of those already in position. As to the actual condition of the colonies, no measure of any kind will at this stage have any bearing on individual colony strength and the possible performance.

The primary factors which determine the outcome — apart from the weather, over which we have no control — are firstly, the race or strain of bee. The former native bee was outstandingly good on the heather, considering the relatively small colonies it formed. However, this bee could not possibly produce the crops of heather honey we now expect from the more prolific strains. The French black bee, a close relation of our old English variety, proved superior to the latter due to her greater fecundity. Unfortunately, the genuine French bee can be truly ferocious, particularly when working on the heather. Indeed, for some unknown reason the most gentle of bees will at times develop a marked hostility on the heather.

Our own strain has over the years given eminently satisfactory results in regard to the heather, when provided with a hive of a corresponding capacity. The view held at one time, that a colony in a large brood chamber would store the greater part of the heather crop next to the brood and little or none in the supers has proved utterly false. As a matter of fact, we often would wish more were stored in the brood chamber than is actually the case. Really powerful colonies tend to store by far the greater part of heather honey they gather in the supers.

52

The two all-important factors to success on the Moor are then, strain and colonies of surpassing strength. A few examples will make the point clear. In 1933 two bee-keepers set up their hives within range of our own bees; in one case merely a few hundred yards from ours. Both used BS equipment and bees of nondescript origin. 1933 proved an outstandingly good season and at its conclusion I had an opportunity to ascertain the results these bee-keepers secured. Both reported an average of 27 lb. and in their view the heather yielded well early in August, but failed when the fine weather returned during the latter part of the month. The flow from 25th-28th August, with a daily net gain approaching 20 lb., resulting in an average of 95½ lb. of surplus per colony, was in fact one of the best I can recollect. There is no doubt, their failure — or relative failure — was due to the fact that these two bee-keepers had the wrong type of bees and the wrong kind of hives. Their colonies were probably composed of mainly old bees, which wore themselves out on the early flow. And as is usual with colonies of this sort, very little brood was raised after their arrival on the Moor. The heavy winter losses reported the following spring confirmed this assumption.

The value of superlative strong colonies was brought home to me in a still more convincing way. Over a period of about fifteen years I had an opportunity to compare the results a commercial beekeeper secured, whose colonies were accommodated in brood chambers holding twelve combs of BS size. The bees in this case were mainly of our own strain. His crops of heather honey regularly averaged about half that of our colonies. Only the relative difference in the actual colony strength could in this instance have accounted for a 50 per cent variation in the respective yields. There was, indeed, no other explanation, a fact unreservedly acknowledged by the bee-keeper concerned.

When conditions are just right the ling will secrete nectar in unparalleled abundance. It would, however, be quite wrong to presume that a given area will support an unlimited number of colonies. By putting more colonies in an area already fully stocked everyone will be injured by such a course of action. It is always possible to find a site for a small number of colonies, but a much more difficult proposition to discover one for a large number.

Courtesy joined to self-interest will be abundantly repaid in instances of this kind. We find that about forty colonies within a radius of a mile is the utmost number an area with an abundance of heather will profitably support in an average season.

As I have already pointed out, on Dartmoor the main flowering period of the ling extends from about mid-August to 5th September. There is often still an abundance of heather in bloom after this date, but will rarely if ever secrete freely thereafter. We have found there is nothing gained by leaving the bees any longer on the Moor. Indeed, much is set at risk by delaying their return to their permanent sites. Moreover, the weather at this time tends to deteriorate rapidly, rendering the return transport a progressively more arduous task. So on or about 8th September we hasten to place all the supers on the escapes, whether full or empty. Two or three days later they are taken home. The colonies are then ready for the return to their permanent sites. We make every effort to get them back as quickly as at all possible, for the sooner they can settle down for winter the better.

3. AUTUMN AND WINTER

a. The closing down

The procedure of bringing the hives back is identical to the one when they are transported to the Moor, except that they return without the supers. The colonies that are brought back each morning are given approximately a gallon of syrup the same afternoon. This quantity of syrup is fed to each colony, irrespective of the amount of stores it may possess. The risk of dysentery, when wintered exclusively on heather honey, seems apparently a universal experience. The syrup now fed is mostly stored in the centre of the brood-nest and consequently consumed first in the course of the winter. The danger of dysentery is thus largely avoided, though by no means completely. After the feeding all colonies are weighed. Those which do not possess a certain minimum weight of stores will receive a corresponding additional allowance of syrup. However, in seasons of total failure, when they return from the Moor virtually destitute, only a certain minimum is fed to ensure their survival of the winter. To feed enough to carry the colonies through until mid-April would, as experience has demonstrated, exhaust the bees prematurely, leading to disastrous results.

Little more can be done on conclusion of the feeding. If a colony is observed to be queenless, when the feeders are removed, it will be given another queen. But an actual check of all colonies for queenrightness is no longer possible, owing to the danger of robbing and the upset an inspection is liable to cause at this season of the year. All our colonies are usually wintered on ten combs. A further contraction is of no beneficial advantage.

The high humidity in winter, particularly in South Devon, and the consequent abnormal condensation in the hives, is one of the great problems of our climate. In an endeavour to mitigate this trouble and to reduce it to an acceptable degree, we experimented with a number of counter measures. A slight through ventilation of the brood chamber seems to offer the most satisfactory results. A draught must be clearly avoided. We have secured the best results by placing a ⅛ in. thick strip of wood between the crown board and brood chamber, over the lugs of the frames at the front and back, allowing the moisture to escape by the slight opening thus formed on both sides of the brood chamber. At this point the moisture is carried away without exposing the cluster to a direct draught.

b. Protection during winter

Although we now and again have to put up with exceptionally severe winters even here in the south-west, we do not provide our colonies with any additional protection. We know that cold, even severe cold, does not harm colonies that are in good health. Indeed, cold seems to have a decided beneficial effect on bees.

About 60 years ago Dr E. F. Philips and G. S. Demuth, who were at that time in charge of the U.S.A. experimental work on bees, advocated a special type of protective wintering case for the accommodation of four colonies. Four hives were set side by side in these cases, forming a cubical block, and enveloped in 4 in. of packing under the hives, 6 in. on the outer sides, and 8 in. or more on top above the crown boards. The claims put forward for this form of wintering seemed to call for a practical test on our part. We accordingly constructed two of these wintering cases, and awaited the results full of high hope.

On the first examination in spring the eight hives were found to be bone-dry and without a trace of mould on any of the combs. So far all expectations were fulfilled, but a great disappointment awaited us. These colonies, without exception, failed to build up. The normal brood-rearing urge, manifested by the other colonies not thus protected, as well as the upsurge of energy and industry, was completely lacking. The colonies wintered in the makeshift hives with little or no special protection, made rapid strides in the spring build-up. Notwithstanding these disappointing results we gave these winter cases a further trial the following winter, but with no better luck.

A few years later Mr A. W. Gale felt likewise compelled to give this form of wintering a trial. Notwithstanding the dissuasion on my part, he had forty cases made. We simultaneously put ours in use again. The trial was now conducted on a total of 168 colonies and in two divergent localities. The outcome proved absolutely identical to the tests made in the first instance. In short: this form of wintering did not only prove a complete failure, but in actual fact had a detrimental effect on the well-being of the colonies. It might well be assumed that in much colder parts of the world this form of protection would prove really satisfactory. This is seemingly not so, for in the course of time this form of wintering was gradually abandoned in the U.S.A. as well as in Canada.

Notwithstanding the abject outcome of these trials, we did not regret having made them, for they revealed important and far-reaching considerations of a kind diametrically at variance to the commonly accepted views and assumptions held on the value of winter protection. The results secured here in Devon as well as in Wiltshire palpably demonstrated that undue protection has a positive harmful effect and that cold – even severe cold – exerts a beneficial influence on the well-being of a colony. As a matter of fact, bee-keepers on the Continent, where extra winter protection was until recently considered essential, have gradually come to the same conclusion as our findings made half a century ago. We have to admit, we are here up against physiological reactions and influences, of which we have little or no knowledge, but which have a decisive bearing on the seasonal development and well being of a colony. Winter losses are not the direct result of exposure to low temperature, but are generally due to a lack of timely cleansing flights, unsatisfactory stores, queenlessness, disease, etc.

A sunny aspect and shelter from prevailing winds are undoubtedly most desirable and beneficial, not only in winter, but at all times of the year. Extra protection from cold may be of advantage in March and April, during the critical period of the build-up in spring. But strong, healthy colonies will manage perfectly well even in adverse climatic conditions. The honeybee is doubtless a creature of the sun, but one that does not need any pampering.

4. ROUTINE PROCEDURES

a. Sterilisation of hives

We cannot hope to secure the most satisfactory results in bee-keeping without a due regard to the hygienic requirements of the honey bee when accommodated in a modern hive. In primitive hives such needs were automatically attended to. The skeps were periodically fumigated over the sulphur-pit and the combs destroyed. There is little doubt, the modern bee-keeper often considers the hygienic needs of his bees as tasks of supererogation. The harm resulting from such a neglect will be conveniently attributed to other causes, as is so often done in bee-keeping.

The periodical renewal of brood combs has already been considered. I am here concerned with the sterilisation of the hives and details of the installation for dealing with this task efficiently and with a minimum of effort. The hives – excluding the supers – are sterilised in a four-yearly rotation. The floor boards receive a scrubbing every spring, but both the floor boards and brood chambers are treated in a caustic soda solution every four years and are then repaired and repainted.

We have an oil-fired automatically operated steam-generating plant and a 30 in. diameter steam pan to facilitate the sterilisation. The steam pan is designed to take two brood chambers at one boiling. They are held in a steel cage, which is immersed and withdrawn from the caustic soda solution with a rope and pulley. One pound of caustic soda is used to 25 gallons of water. After the sterilisation they are scrubbed and rinsed in clean hot water.

Most bee-keepers nowadays creosote their hives. We did so too at one time, but did not find it satisfactory. To keep the hives in reasonably good condition they would need an application at least every two years. In our heavy rainfall creosote is washed away very quickly. On the other hand, an oil paint is quite useless for single-walled hives. The paint comes off in blisters, due to the condensation formed inside the hives. For many years we made use of an outdoor quality distemper, which allowed the moisture to escape, and which we found in every way superior to creosote. Only two applications were needed and the cost was very much less than that of a good quality oil paint. For some years now we have used an emulsion paint, which also requires only two applications and which we find still more durable than the distemper. Where the use of an emulsion paint is contemplated, care should be taken that the particular brand does not incorporate an insecticide. On the external parts of the bottom boards an enamel paint is used; so also for the roofs. And to render the appearance of the hives more attractive – at no extra cost – the floor and roof is painted in a shade of salmon, the brood chamber in cream, the supers are stained brown.

At one time a whitish fungus made its appearance in the interior of the brood chambers, causing the affected parts to perish within a year or two. An application of Cuprinol, whenever the hives are repainted, has completely obviated this trouble. The interior of the floor boards are also treated periodically with Cuprinol – the clear variety is used in both instances. The parts treated with Cuprinol must, however, be well aired before use as otherwise robbing may be caused by the odour.

The steam-generating plant and installation has also proved invaluable – indeed indispensable – for the sterilisation of the brood and shallow frames. Of the former about 1,200 need sterilisation annually; of the latter no less than 6,000 after a good crop of heather honey. All the frames are boiled in caustic soda and subsequently rinsed in clean water. After the rinsing they are as clean and free of wax and propolis as when they were new. The steel cage takes about thirty brood frames or fifty of the shallow type at one boiling.

b. Wax rendering

The steam installation comes likewise most useful for the wax rendering operations. Indeed, steam permits of an instant control and is therefore the only really safe way of dealing with beeswax and the rendering of old combs. In a good season our crop will amount to close on a ton. An efficient installation is therefore very necessary. In the case of old brood combs we are able to extract 50 per cent over and above the actual weight of the foundation used; shallow combs will yield double the amount.

The wax is cast in moulds holding approximately 1 cwt. The moulds are made of plywood and lined with tinned copper plate. As the wax cools, which takes abour four days, it contracts allowing the blocks to slip out of the moulds freely.

Spring 1910 At that time we did not possess two hives of identical design; the WBC hive proved a first step in modernisation

By may 2015 a change-over ti the Burgess Perfection hive was in progress

From 1922 our hives were set out in groups, which we found virtually eliminated drifting

On the Moor in August 1920. The 100 hives were according to custom set out in rows, all facing in a southerly direction. The drifting caused can be readily ascertained by the number of supers per hive

The home apiary in 1938, by now completely modernised

Notwithstanding the flimsy construction of the makeshift hives, the crop secured proved satisfactory

One of our out-apiaries. Where the terrain permits, as here on a side of a hill, the hives are set out in groups of two

Every out-apiary is provided with a hut for storing supers, etc

Our first task in spring consists of exchanging the floor boards for clean ones. They are constructed with rabbets on three sides and a drop of 1 in. from back to front to prevent water from collecting inside the hive

The tray-feeder. Bees gain access to the syrup inside the tin cover which fits over the central feeder block

The supers hold ten wide shallow combs and the spacing is achieved by a series of notches, eliminating any projections

Above left: A hive ready for transportation, with screen, entrance slip and the fastening rod at the back in position.

Above right: At the conclusion of the season and before the final removal of the supers. Each super when full holds 50lb. of honey.

A load of 40 hives, complete with roofs and hive-stands, ready for the return trip from the Moor

Above: The honey storage tanks – 11 in number with a total capacity of 271/2 tons.
The automatic bottling machine will fill up to 2,000 1lb. containers in an hour

Below left: Each storage tank is fitted with a hot water coil, connected to a gas heated
thermostatically controlled boiler, for liquefying the honey when granulated.

Below right : The installation for preparing the syrup by the cold water process.
The tank holds about 1½ tons of syrup

Above: The boiling pan in which the sterilisation of hives and frames and the rendering of wax is performed. The steel cage, in which the hives and frames are held for sterilisation, is shown suspended over the boiling pan: the rinsing trough is to the left; a wax mould to the right

Below: The oil-fired, automatically operated steam boiler – an essential component to facilitate the sterilisation of hives and rendering of wax

We place great value on shelter and good drainage. From this point of view the out-apiary here shown is ideally positioned

On Dartmoor in 1927

Part of the home apiary showing the four-hive layout and the queen-rearing house

Interior of the queen-rearing house. On the right, the heating cabinet for the queen-cups, grafting table and a hive on scales: on the left, an electrically-heated incubator with a capacity of 1,000 queen cells.

Above: The mating station on Dartmoor, established in 1925 and in continuous use since then

Centre: A close up of the main mating station

Below: A supplementary mating station in another part of the Moor – used for experimental crossings

Above: A close-up of the interior of our mating hive for four nuclei, each on four Dadant half-size combs

Centre: The mating hive we first used. This also accommodates four nuclei, but on three half-size BS combs

Below: The tray-feeder adapted to give four nuclei simultaneously access to the syrup. A cut-out in one of the tin caps permits one of the nuclei to clean the tray when empty

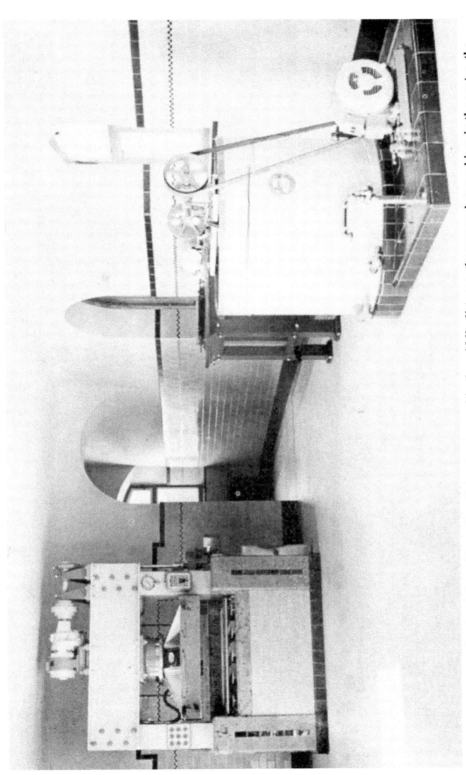

The extracting plant. On the right, a 44-comb radial extractor, and in the middle distance, the uncapping machine: in the centre, the hydraulic heather honey press. This press will handle two tons of honey per day, with a loss of no more than 1.2 per cent of the crop

PART III
BREEDING AND
RAISING
OF QUEENS

1. BREEDING

a. The importance of breeding

As the reader will have observed all along, breeding and strain form the cornerstone of the bee-keeping as practised at Buckfast. Up to now I have discussed the factors and preconditions necessary to ensure the unfolding of a maximum honey gathering ability of every colony, the ultimate aim of every endeavour in bee-keeping. In breeding we deal with the hereditary predispositions of a desirable and undesirable nature. The former we endeavour to foster and to intensify ; the latter we try to minimise and to eliminate if possible. In the more advanced sphere of breeding our aim is to form new combinations, new strains. This last method of breeding places us in a position to utilise the immense fund of valuable qualities which Nature has put at our disposal in the various geographical races of the honeybee.

To set the present-day efforts in the right perspective we have to bear in mind that from the beginning of time and up to a hundred years ago, the breeding of the honeybee was the exclusive prerogative of Nature. Man had no influence in this direction before the introduction of the movable-frame hive. But Nature had no interest in the attainment of a high average honey yield per colony. All her efforts in breeding were concentrated on the maintenance and spread of the species. In the various races she has handed down to us the raw material for the advancements that await us in the future.

We have at all times set the utmost importance on breeding, realising that breeding alone can open up the possibilities of a permanent and progressive improvement in the honeybee and enhance economic returns from bee-keeping. This holds good in the breeding of all livestock and plants of economic value. In the case of the honeybee breeding will doubtless play a role of even more importance – it will in fact determine the actual progress bee-keeping will achieve in the future. However, as experience clearly has demonstrated, in order to attain our aims, the task facing us must be founded on a solid and realistic basis. Enthusiasm alone will not suffice.

In the following pages I shall confine myself mainly to the strictly practical aspects of breeding. However, references to certain specific problems are unavoidable, for they form essential considerations underlying our endeavours and the practical results achieved. Though they may seem of an academic nature, they are in fact of vital practical importance. Every well-informed bee-keeper, whether professional or amateur, should be conversant with the things that matter in the breeding of the honeybee.

b. Pure stock or hybrids

Bee-keepers have the choice of two alternatives: the breeding of pure stock, or the use of pure stock for the creation of hybrids. A corresponding selection is the key to success in either case. Nature relies on selection to attain her ends. She is indeed ruthless; her selection is based on the survival of the fittest. We can, however, gather some hints and draw some useful conclusions from the means she uses to attain her objectives. We shall discover that the mode of propagation and multiplication she designed ensure a constant crossing and interchange of hereditary factors. Multiple mating is doubtless one of her most effective ways of achieving this aim. The never-ending interchange, resulting from one queen mating to possibly ten or more drones, seemingly constitutes the norm in the propagation of the honeybee. The mating at a distance, up to five miles from the parent hive, is a further contributing factor, as too is swarming.

c. Line breeding

There is no doubt whatever that the bee-keeper, who strives to develop a highly uniform strain of bee – under the assumption that uniformity will *ipso facto* denote a high uniformity in performance – contravenes the most elementary dictates of Nature, for uniformity can only be attained by close inbreeding. As we have seen, Nature abhors inbreeding and endeavours to prevent it by every means at her disposal. On the other hand, we also know there can be no success in the breeding of the honeybee without going in some measure against Nature. But certain limitations Nature has set must be observed, if failures and disappointments are to be avoided. As can be concluded from Nature's ways: the honeybee is extremely susceptible to inbreeding – a fact not widely recognised until recently, but which we were fully aware of more than forty years ago. An intensification of this or that quality, the aim and purpose of the development of pure stock, can only be attained by a corresponding measure of inbreeding which, unless effected on a broad basis, will inevitably result in a loss of vitality. The loss will be manifested in diverse and often obscure and insidious ways. Where inbreeding is carried beyond a certain point, the total ruination of a strain will be the inevitable outcome.

Whereas the development and use of pure stock is almost universally regarded as an end in itself, we at Buckfast consider it essentially as but a necessary step in the production of specific crosses. Only thus can we hope to secure the full advantages pure-bred stock can offer.

d. Cross-breeding

If we look around we shall note that all the worthwhile improvements in agriculture, in the breeding of livestock or plants, have been brought about by cross-breeding. Indeed, the present-day level of production in all branches of agriculture would be inconceivable without the help of cross-breeding, which is unquestionably the key to all worthwhile progress in the improvement of domestic stock and plants. The honeybee is no exception. If our experience is anything to go by, bee-keeping, whether carried out for pleasure or profit, can less easily afford to deny itself the advantages offered by this form of breeding than other sections of agriculture.

Pure-bred stock will always occupy an important role in beekeeping, but very often far too much is expected from it than objective reflection and practical experience warrant. It will fail – must necessarily fail – where a substantial increase in performance is hoped for. Cross-bred stock, on the contrary, is able to ensure higher yields with almost unfailing certainty.

In cross-breeding the honeybee we are, however, confronted with a number of unusual problems, and also often faced with some unusual and unexpected results. For instance, reciprocal crosses are rarely identical. In matching gentleness with gentleness we do not invariably secure greater gentleness, or even a gentleness equal to that of the parent stock: on occasion we may actually obtain bad temper. But the reverse is also true, namely, that bad-tempered parentage may give rise to exceptionally gentle offspring. The many unexpected results secured in cross-breeding are doubtless in part responsible for the confusion of views on its value to bee-keeping.

I will deal first with the unusual phenomena of the difference between the maternal and paternal influence. We know from practical experience that with few exceptions the queen exerts the dominant influence on her offspring. This holds good in line-breeding and cross-breeding. One of the most classical examples of this dominance known to me is the transmission of resistance and susceptibility to Acarine. With queens of highly resistant stock, the power of resistance is unaffected in the first generation by her mating. Needless to say, if the drones originate from a susceptible line, the susceptibility will inevitably show up in later generations. The same holds good when a queen transmits a high susceptibility. This dominance in the maternal heredity is of great practical importance in cross-breeding the honeybee. It enables us to bypass or to modify certain undesirable characteristics in their harshest form. However, maternal dominance does not hold good for every characteristic and, indeed, is not commonly manifested in the unequivocal manner as in the case of Acarine. We have, for instance, found that in Syrian queens crossed with Buckfast drones, the extreme aggressiveness of the Syrian is greatly modified; here the paternal influence dominated over the most extreme form of irritability.

From this case just cited the conclusion might well be drawn that, with regard to temper, the drone has the dominant influence, confirming an assumption very widely held. However, as already pointed out, gentleness matched with gentleness does not invariably beget greater gentleness, but will occasionally give rise to bad temper. For instance, the Caucasian is universally regarded as the most gentle of all races, yet when mated to Italian drones the resultant offspring is liable to be anything but good-tempered. The same holds true for Carniolans crossed with Buckfast drones and equally for the reciprocal mating. Indeed, crosses from a particular strain of Carniolan, widely favoured throughout central Europe, have often turned out quite unmanageable.

Fecundity is also a characteristic which is subject to variations in a first cross, though it is commonly held that a F-1 queen is invariably exceptionally prolific. According to our findings, hybrid-vigour, or heterosis, has no influence on fecundity in a first cross Carniolan-Buckfast, Carniolan-Italian, and Carniolan-Greek. The brood pattern is admittedly more compact, but there is no significant increase in the first generation in either the brood area or egg-laying rate. A greatly enhanced fecundity will, however, manifest itself in the subsequent generations. On the other hand, the reciprocal crosses Buckfast-Carniolan, Italian-Carniolan, and particularly Greek-Carniolan, produce the opposite results in a first cross, namely, a greatly increased fecundity. A cross between Cyprian-Buckfast, Anatolian-Buckfast, and especially Greek-Buckfast, are classical examples of a marked accentuation of fecundity in an F-1 generation. The Anatolian-Buckfast cross is rather unusual, for pure Anatolian queens are far from prolific, judged by present-day standards.

I could cite many instances of contrarinesses met with in the first hybrid generations. One thing seems certain: in cross-breeding the honeybee we are rarely in a position to foretell specific results with any measure of certainty. Only experience can give an indication of the vagaries a particular cross is likely to manifest.

Apart from the combination of a particular set of characteristics secured, whenever two distinct races are crossed, an additional quality of great economic value is gained, viz. hybrid vigour, also known as heterosis. Indeed, if it were not for heterosis, hybridisation would lose much of its economic importance. Heterosis brings about the results opposite to those of inbreeding; it enhances in varying degree the general vitality, health, growth and productiveness. This holds equally good in the honeybee. As a matter of fact, its high susceptibility to inbreeding, and the clear tendency on the part of Nature to avoid matings between closely related individuals – the closest forms of inbreeding, as sometimes applied in other cases, are not even possible in bees – would indicate that the honeybee is a particularly suitable subject for hybridisation and the utilisation of heterosis. This is in fact the case, as our experience and the accumulated evidence and the positive results from controlled matings have demonstrated.

e. The influence of heterosis

The unfavourable views held on the economic value of hybridisation are, I believe, based on an incomplete appreciation of the unusual reactions heterosis brings into play in the honeybee. The common run of random matings and the resulting crosses have admittedly little to recommend them, although there are exceptions and now and again we meet instances of surpassing performance. But these are the result of sheer luck and chance and, unfortunately, fortuitous results cannot usually be repeated at will. However, our controlled hybridisation trials have revealed that in the honeybee the most satisfactory economic returns are often not obtained in a first cross, as in the breeding of domestic animals and plants, but in the following hybrid generations. This diversion from a universally accepted rule can be readily explained, although to my knowledge no one has up to now attempted to do so.

It will be appreciated that heterosis enhances not only the desirable qualities and dispositions but also the undesirable ones. Among the latter we have to class the swarming tendency. A basic instinct such as this tends to dominate over all the other qualities, with the result that a first cross often dissipates its exceptional abilities in uncontrollable swarming. However, this unbounded swarming propensity is greatly modified in the F-2, thereby giving full scope to the qualities of economic importance. In swarming we are obviously dealing with an hereditary disposition exclusively confined to the honeybee.

The possible uneconomic consequences of heterosis and its bearing on swarming will be perhaps most convincingly demonstrated by an example from practical experience. At one time we tested a certain first cross – queens of a well-known Swiss strain crossed with drones of our own. In the year in question our average per colony amounted to 45 lb.; that of this particular first cross to no more than 22 lb. The huge disparity in the yield was almost entirely due to the overwhelming swarming propensity of this particular cross, notwithstanding the fact that the maternal stock had a high reputation for non-swarming in its country of origin. This example shows that the economic difference, resulting from an extreme swarming propensity, can be of a serious magnitude. We have therefore at Buckfast made it a rule, when dealing with unknown stock, to test such crosses on a small-scale initially. This also applies when we know from experience that heterosis will in a particular case accentuate the swarming instinct unduly. Moreover, in instances of this kind nothing seems to be gained by a wide selection based on a large number of colonies in a first cross.

According to our findings the following are the most notable examples of this category, which in the following generation however — either in a back cross or mated *inter se* — give results of outstanding economic value: French queens x Buckfast drones; Swiss x Buckfast; Tellian x Buckfast; and Carniolan x Buckfast. We have evidence that if drones of other races or strains had been used, the swarming impulse would have been still further acentuated in each of these crosses.

A number of first crosses take up an intermediate position. In these heterosis has a decided bearing on swarming, but productivity is not adversely affected as in the cases just cited. In this class I would put Cyprian queens x Buckfast; Syrian x Buckfast; and Caucasian x Buckfast.

There are exceptions to the two foregoing groups in which

heterosis does not materially affect the swarming tendency, and where we can therefore reap the maximum benefits from heterosis in a first cross. According to our findings these are: Anatolian x Buckfast; Buckfast x Carniolan or Greek; Greek x Buckfast or Carniolan.

Only the latter crosses are in my estimation of economic importance to the majority of bee-keepers. I should perhaps point out, while our experimental work is primarily based on the Buckfast strain, somewhat similar results will in all probability be secured by the substitution of queens or drones of Italian origin.

I have briefly dealt with the problem of hybridisation from the point of view of the practical bee-keeper and indicated how to utilise heterosis to the best advantage. But cross-breeding has very much more to offer. It presents the means by which we are empowered to form new combinations, new races synthetically. In creating new combinations the breeder produces something of permanent value. The full economic benefits of a new combination will in turn come to light in a renewed and accentuated heterosis, for the more productive the parental stock utilised in the formation of a cross, the more marked the heterosis. Furthermore, each new combination should prove a step forward in the overall advancement in breeding and in the economic prospects of bee-keeping.

I should perhaps emphasise: the crosses here indicated were in every instance the result of controlled matings. Our strain formed the basis of these experiments and at the same time the standard or yardstick by which the respective results were evaluated. While only a relatively small number of crosses have been cited, these breeding experiments embraced in fact almost every known race of the honeybee.

I laid stress on the uncertainty of foretelling the results in cross breeding and pointed out that gentleness matched with gentleness does not invariably beget a greater gentleness. But the opposite is equally true. In 1938 we secured a surpassingly good tempered line which we developed from a French-Buckfast cross. This line proved in fact far gentler than any other strain we have ever come across and it was also quite dissimilar in every other way to the original parental stock. In colour it was a deep golden and in its economic characteristics as near perfect as we could possibly desire. It was, unfortunately, afflicted with one major defect: an extreme susceptibility to Acarine derived from the French parentage. This susceptibility was so intense as to render the line economically valueless.

2. OUR AIMS IN BREEDING

I have up to now outlined the main considerations which govern our methods of breeding and also endeavoured to show that the development of our strain has at no time been regarded by us as an end in itself. Indeed, we consider our strain primarily as a stepping stone to ever more reliable crosses and combinations. It has been our constant endeavour to combine, fix and intensify all the best qualities having a bearing on performance in this strain. But our ultimate aim is the formation of a bee that will give us a constant maximum average crop consonant with a minimum of effort and time on our part.

The aim we have set ourselves includes therefore not only qualities specifically concerned with performance, but likewise all such that bring about the fulfilment of our second stipulation, namely, the reduction to a minimum of the effort and time required in attending a colony. A clear appreciation of the possibilities and limitations we have to deal with to ensure success in our endeavours seems essential. I therefore regard a detailed list of the more important characteristics as desirable at this juncture, including a grouping in regard to their relative importance.

a. Primary qualities

1. <u>Fecundity</u>. An adequate fecundity is a paramount requirement, since without colonies of maximum strength the best crops cannot be gathered. Fecundity alone is, of course, not the decisive factor, but it is the essential basis of exceptional performance. A queen which at a certain point in the colony's development cannot fill eight or nine Dadant combs with brood has not the

necessary qualifications for our purpose. Moreover, this extension of the brood area must be attained and maintained spontaneously, without any artificial stimulation.

I am aware these stipulations are widely at variance with the views often expressed. The contention: we want honey not bees is a palpable absurdity. Of course, we do not want bees which turn every pound of honey gathered into brood. There are, admittedly, such strains. But it is the bees that gather the honey and the more powerful a colony, of the right type of bees, the greater the honey-gathering potential of such colonies. However, one desirable characteristic necessitates and presupposes a whole series of other qualities. We cannot really consider one particular characteristic in isolation. There is an interdependence between one particular trait and many others, just as in a chain one link is dependent on another, and one member by itself cannot perform its allotted role. The fulfilment of our overall objective is, indeed, determined by the extent the various characteristics mutually complement each other.

I should perhaps in this connection point out that fecundity is subject to two factors: on the physical ability of a queen and a disposition on the part of the bees or brood-rearing incentive. As we know, in some strains the honey-gathering incentive will dominate over the brood-rearing incentive; in the over prolific strains the reverse is the case.

2. Industry. Of all the qualities, a boundless capacity for work is doubtless the foremost requirement. Industry is the lever which transforms all the qualities of economic value to our advantage. The honeybee is universally regarded as a symbol of industry, yet we know there are strains which are good-for-nothing but also others that are endowed with an inexhaustible industry. As our comparative tests have shown there are also great differences between the various races. Industry is doubtless based on a series of hereditary dispositions of a cumulative nature.

3. Resistance to disease. There can obviously be no maximum performance where disease of one kind or another undermines the vitality of a colony. Sound health is the *sine qua non* of

successful bee-keeping. The task of developing strains resistant to the various diseases is therefore one to which the utmost importance must be attached. Very strangely little or nothing has been done in this connection in the breeding of bees, notwithstanding the great potentialities, as our own experience has clearly demonstrated. Admittedly an undertaking of this kind presupposes conditions and facilities of a type not ordinarily available. We must, however, make a sharp dinstinction between resistance and immunity, and realise that between extreme susceptibility and immunity every degree of resistance will usually be manifested. So far as I have been able to ascertain, the possibility of an immunity can only be entertained in the case of Paralysis and Sacbrood. A high resistance to Acarine, to an extent as to render this disease of no practical consequence, can be readily achieved. In the case of Nosema, the degree of vitality seems the deciding factor and inbreeding the primary predisposing cause.

4. Disinclination to swarm. A highly developed disinclination to swarm is doubtless an indispensable prerequisite in modern bee-keeping; when bees were kept in skeps the primitive way, the reverse was the case, for swarming was then the only means colonies could be propagated and multiplied. In modern bee-keeping swarming not only causes untold extra work and loss of time on the part of the bee-keeper, but is also often responsible for a serious reduction in the amount of honey a colony produces. Indeed, a race or strain of bees endowed with every desirable trait but given to swarming, will from the strictly practical point of view, prove of little value, for all the qualities of economic importance will be dissipated in idle swarming. The case of the Swiss cross already cited is a classic instance.

While a truly non-swarming bee is almost certainly beyond our reach, strains which will normally only swarm in exceptional circumstances are undoubtedly possible, are in fact already at our disposal. Further progress in the direction of a non-swarming strain seems merely a matter of time and perseverence.

b. Secondary qualities

Fertility, industry, resistance to disease and disinclination to swarm are in my estimation the basic qualities of economic importance and form our primary aim in breeding. The characteristics I am now going to outline are not essential in the same way, but they are of great importance as each contributes its respective share to an intensification of the honey-gathering ability of a colony. While we necessarily have to consider each trait and quality more or less in isolation we must never lose sight of the fact of their interdependence and interaction and their cumulative effect on performance. On the other hand, one major defect can also render an entire combination of desirable characteristics valueless.

1. Longevity. This quality must in my view head the list in this section. Presumably no one will question the fact that substantial differences in longevity exist. It is a hereditary trait which, as our experience would indicate, promises major possibilities in breeding. A prolongation of the lifespan of a bee, even if only by a few days, will denote a corresponding increase in the effective foraging force and capacity of a colony. This in turn will means a higher colony performance at no extra cost.

So far as I have been able to determine longevity is based on two factors: on heredity and on the quality and abundance of the food provided during the brief larval period of a bee. We are here dealing with two distinct hereditary dispositions. Certain races, notably the Anatolian, Carniolan and the dark west-European varieties are long-lived, a fact which is also manifested in the longevity of the queens. A life-span of up to five years in the case of queens of these races is no uncommon experience. On the other hand, there seems a relation between fecundity and longevity. Ultra-prolific strains – with possibly one exception – are invariably short-lived; exceptional longevity is, on the contrary, found among the moderately prolific strains. Our former native bee was probably one of the most outstanding examples of longevity.

2. Wing-power. The ability to forage beyond a commonly accepted flight range can prove a material factor in the performance of a colony. Here again I have observed substantial differences. The most striking example was the native British bee. Up to the time of its extinction we regularly secured a crop of heather honey from the colonies situated in the Abbey grounds. The nearest heather is 2¼ miles from Buckfast and some 1,200 ft. higher than the Abbey. In spite of the distance and climb involved the native colonies and crosses managed in 1915 to gather close to 90 lb. on an average, including winter stores.

3. Keen sense of smell. Presumably an exceptional wing-power presupposes a keen sense of smell. Without such a keen sense a bee would not be tempted to forage beyond a certain range. This trait has, however, its drawbacks, for it tends to lead to robbing. Indeed, my observations have led me to believe that the two traits are complementary. Colonies of outstanding performance seem invariably first on the scene when there is an opportunity for robbing.

4. Instinct of defence. The most effective remedy against robbing is an acute instinct of self-defence. A highly developed instinct of defence is an essential requirement in a modern apiary, where many colonies are congregated and hives have to be opened in times of scarcity. This trait is most highly developed in the races of the Middle East, where the struggle against innumerable enemies, enemies of a kind which the bees in temperate zones do not have to face, has doubtless been responsible for the high development and intensification of this instinct.

5. Hardiness and wintering ability. Hardiness in the honey-bee comprises a number of aspects. It is really not so much a question of resistance to extreme cold, but rather an ability to winter on stores of inferior quality for long periods without a cleansing flight, as well as a reaction to sharp changes in temperature, disturbances, etc. For instance, the Carniolan will fly in winter at the slightest rise in temperature in excess of 40°F, but to no useful purpose. Colonies of our own strain will in identical circum-

stances remain perfectly quiet – in fact behave as if dead from the beginning of November until an opportunity occurs for a good cleansing flight at the end of February or early in March. Every undue activity in inclement conditions, in winter and spring, will entail a useless waste of energy and loss of bees.

6. <u>Spring development</u>. The manner colonies will build-up in spring is within certain limits determined by heredity. Some races, notably the Carniolan, start raising brood prematurely, before climatic conditions are really propitious. Colonies thus disposed fritter away their vitality in flights and endeavours which serve little or no useful purpose and, as experience has amply demonstrated, such forwardness often leads to severe set-backs if prolonged spells of bad weather supervene. Moreover, the loss of stamina renders colonies of this kind highly susceptible to Nosema.

We favour strains that delay the build-up until a time in spring when more settled conditions prevail, when the bees can go foraging to good purpose. In most instances, these late starters build-up in giant strides and surpass in strength the precocious type. The most notable example among the late starters and rapid build-up is undoubtedly the Anatolian. Indeed, this race has set an ideal in regard to this particular characteristic unmatched by any other variety.

Linked to this disposition must be the equally important one: once the spring development has begun it must go forward irrespective of changes in the weather, spontaneously and without the need for any artifical stimulation.

7. <u>Thrift</u>. Frugality or thrift is a quality closely connected with the seasonal development of colonies. Here again we have wide differences between one race and another. I regard the Anatolian the classic example of thrift; the American Italian strains as cases of extreme extravagance. The lack of an appropriate measure of thrift seems to me one of the serious deficiencies of many of the present-day strains. It is one to which we have been giving much attention in breeding in the course of the years.

8. Instinct of self-provisioning. In a well-bred strain this characteristic will normally manifest itself towards the close of the season, thereby ensuring an adequate supply of winter stores. Some races and strains will manifest this disposition almost at any time of the season. In such cases it can prove a serious drawback, for it will effectively eliminate the force of bees needed for gathering a crop at a later period of the season. The Carniolan is probably most given to premature brood chamber storage; the Italian forms the other extreme. The requirements of modern beekeeping demand a middle course between the two extremes. The actual amount of stores a colony possesses on returning from the heather plays an important role in our evaluations and in the selection of queens destined for breeding purposes.

9. Comb building. This is another important characteristic with a bearing on a number of other traits, particularly on swarming. A keenness to build comb seems to increase the zest for every form of activity of economic value. Indeed, one can always tell when a colony is actively engaged in drawing out foundation that everything is in order.

Among all the races I know, the old British native proved the most outstanding comb builder. She somehow managed to draw out foundation with an extraordinary rapidity and to a perfection almost unequalled by any other race. We have been able to retain this valuable trait to a considerable extent in our strain.

Closely connected with this question of comb building is the inclination to construct drone comb. The construction of drone comb and the rearing of drones beyond a certain measure is clearly uneconomic. Hybrids, as a rule, tend to extremes in this direction, particularly when any feeding is done, less so when they have to work for a living. But much can be accomplished in mitigating this tendency by selection.

10. Gathering of pollen. The urge to gather pollen is not the same as the urge to collect nectar. The Italians are seldom found gathering an excess of pollen; likewise the Carniolan. The western European group of races, particularly the French variety, can all be classed as pollen-hamsters. The French bee will carry

pollen through the excluder and store it in the supers. This phen-omenal urge is hereditary, and in districts poor in pollen – or where the pollination of crops is a major consideration – this trait would repay cultivation, more especially where there is a scarcity of pollen in the autumn months, for an insufficiency at this period of the season is widely regarded as a primary cause of Nosema.

c. Qualities of indirect value

The characteristics discussed so far have a direct bearing on honey production. Now we can turn to others which do not influence production, but which are essential for the realisation of our secondary objective, namely, the reduction to a minimum of the time and effort involved in the seasonal care and attention demanded to ensure maximum production results per colony. The traits included in this group facilitate the tasks of the bee-keeper; some have likewise an economic and aesthetic value.

1. Good temper. Although bee-keepers are divided in their opinions on the value of most qualities, they are almost in com-plete unanimity on one issue – in their approval of good temper in bees. I came across only one exception and in this case bad temper was of special value in keeping thieves from interfering.

Bad temper renders every task more arduous, slows down the work and causes an unwarranted loss of time, apart from the con-stant danger of unpleasant incidents with neighbours and farmers. Very fortunately gentleness is a hereditary disposition which can be easily bred into a strain. There is in fact no difficulty in breeding sweet-tempered bees in a few generations from a cross of the worst type of 'stingers'. It is sometimes held that there is a connection between temper and performance and that in the transmission of good and bad temper the drone plays the decisive role. This may occasionally be so but does not hold true universally.

Bad temper is often linked with a decided aggressiveness in some races, particularly in the western European varieties. Bees of this kind will sting without any provocation and can prove very dangerous. Indeed, I regard this as a distinctive racial trait of the western European bee. The Middle Eastern races, on the other hand, will not assault anyone except when provoked, but will then attack en masse and at considerable distance from their hives and with an unparalleled ferocity.

The idea of a stingless honeybee is, of course, a pipe-dream. But a bee that will not make use of her sting, or will do so only in quite exceptional circumstances, seems to me a real possibility.

2. Calm behaviour. Bees that stay calm when manipulated will greatly facilitate the work. The nervous type that tumble off the combs or tend to mill around when a colony is checked render the finding of a queen a most difficult and time-consuming task. Highly nervous bees seem particularly susceptible to Nosema apart from any other drawback.

The Carniolan is doubtless the classic example of calm behaviour under manipulation. It is a quality we highly prize from every point of view.

3. Disinclination to propolise. The habit of the majority of races to cover the inside of the hive with a coating of propolis is one of their most unpleasant traits, and one that can substantially increase the work of the bee-keeper. It is a trait we have made every effort to eradicate, but it is a disposition that is bound up with a number of factors. But the Egyptian bee uses no propolis and the Carniolan, at least some of the strains, will use wax in place of the resinous substances. So there is hope that the disposition of using propolis will eventually be eliminated.

4. Freedom from brace comb. The presence of brace comb, which many races and strains construct between top-bars, crown boards, etc. is another most undesirable trait. It renders the inspection and manipulation of combs not only a difficult and arduous task, but will also – if the brace comb is not removed – cause the death of many bees and occasionally that of the queen. The Caucasian displays this undesirable trait to the greatest degree; the Cyprian shows no trace of it. In primitive conditions brace comb doubtless served a purpose whereas in a modern hive

it is an unmitigated nuisance. Happily, it is a trait which can be easily eliminated by careful breeding.

5. The art of making attractive cappings. Good cappings are of special importance where sections are produced. The art of making attractive cappings is a trait which seems tied up with and dependent on many factors. While we have in some measure succeeded in fixing this elusive faculty we have by no means attained the ideal set by our former native bee. Her cappings set a standard of perfection of unsurpassed excellency.

6. Keen sense of orientation. A highly developed sense of orientation helps to ensure bees will return to their own hive thereby avoiding the many drawbacks and risks drifting entails. In countries where from time immemorial colonies are accommodated in close proximity to each other, as still is the practice in the Middle East today, drifting is virtually unknown. In the course of hundreds of thousands of years Nature eliminated individuals of defective orientation. The most outstanding cases of a highly developed sense of orientation is found in the Syrian, Cyprian and Egyptian races; to a lesser degree in the Carniolan and least of all in the Italian.

I have already indicated the undesirable consequences of drifting. The risks involved can be in some measure mitigated by an irregular positioning of the hives. While a highly developed ability of orientation is not as important in such circumstances, it is nevertheless a faculty of much value if an undue loss of queens returning from their mating flights is to be prevented. In the losses thus sustained we have a fairly accurate gauge of determining the ability of orientation manifested by the various races or a particular strain. In our comparative tests the oriental races have clearly come out on top. Indeed, in one instance, at the end of August 1920, when conditions were far from favourable, we had a batch of 110 Cyprian queens of which number only one failed to return. In general the average loss amounts to about 18 per cent.

These in brief are the main qualities we have taken into consideration in the development of our strain and in cross-breeding. There are numerous other traits and genetic dispositions – of a desirable and undesirable kind – but there would be no point in enumerating them here as they are mainly of a scientific rather than practical importance. And, as may be assumed, the honeybee is not free of oddities and anomalies.

Broadly considered, performance is dependent on the interaction of a whole series of hereditary dispositions and the more these factors complement each other, the higher the level of the performance. While we place no particular value on external characteristics and uniformity in colour, they are nevertheless useful indications in the synthesisation of new combinations.

I have again and again drawn attention to the valuable qualities of the former British native bee. There is no denying that this bee was the happy possessor of a number of quite outstanding qualities. Indeed, in all my endeavours I have set myself certain of her characteristics as models of perfection. But as we have seen this holds no less true in regard to many equally important traits possessed by other races. The British native bee had no claim to absolute superiority; she had in fact a great many serious defects. The assumption that an indigenous bee must necessarily prove superior in her native habitat is completely erroneous. Assumptions of this kind are usually based on the supposition that in the course of untold thousands of years natural selection would produce a bee most perfectly adapted to the particular conditions of its environment. Unfortunately, this plausible argument is based on a false premise. As I have already pointed out, Nature nowhere selects for performance as her objective, still less for the highest standard of performance. Nature has at all times and everywhere bred for the maintenance and spread of the particular species. She, furthermore, could not do in any way what the most progressive breeder is able to do, that is, bring out and develop a particular factor not already present in an individual. Nor could Nature bring together the various races, possessing particular valuable characteristics, that have their habitat in widely different parts of the world. She left this task and the realisation of its immense possibilities to the initiative of modern bee-keeping.

d. The importance of comparative tests

In bee breeding we cannot possibly hope to achieve any success of economic importance without reliable comparative tests, without large-scale experiments that provide the kind of data needed. But this type of evaluation is one of the most complex problems in bee breeding. The results obtained are always of a relative nature: relative to this or that race, strain or line; relative to a certain environment and circumstances. The honey flow itself is subject to wide fluctuations from year to year and from one district to another. Even a distance of a few miles can make a substantial difference in regard to the honey flow and every aspect of bee-keeping. Indeed, the results we secure have always to be restricted to a certain year and particular environmental circumstances. There are no two seasons alike.

While we must necessarily take into consideration the particular circumstances in the evaluation of performance and production, mistakes in the assessment usually spring from a different source. Comparisons made within the same line or strain, can determine the relative value of the colonies of the particular line or strain, but a true evaluation cannot be made without wider comparisons. Other strains of the same race – and where possible other races – kept in the same apiary or apiaries at the same time, can alone give us positively reliable comparative results. Moreover, the more comprehensive the scale of such tests, the more reliable the evaluation and the greater the likelihood of success in the ultimate choice.

Our comparative tests are designed to preclude a possibility of failure as far as is practical. Our ten apiaries are in localities with markedly different honey flow conditions. In some the soil is sandy and light, in others of average composition; in others it is of a very clayey nature. In dry seasons we get the best crops of almost pure white clover honey from the localities with heaviest clay, but in very wet seasons they fail us completely. Wintering too on clayey soil is always a problem, because of the excessive dampness. On sandy soils the reverse is the case. Spring development has its peculiarities in each locality. In the valleys where the permanent sites are there is hardly ever much snow, but on Dartmoor

– where the mating station is situated, and where the young queens undergo their initial testing in the nuclei – the climatic conditions are exceptionally harsh in summer and winter.

At all the apiaries special precautions are taken to prevent drifting which, as I have already pointed out, can lead to a false evaluation of the performance of colonies. The queens of a particular line or cross are distributed in about equal number in every apiary, to permit us to ascertain as accurately as possible their respective qualities and abilities. In 1949, for instance, we secured an overall average of 145 lb. of surplus per colony. However, twenty-two colonies, all headed by daughter queens of a par ticular breeder (one of the six breeders bred from the previous year), averaged 185 lb., or 40 lb. more than the other colonies. Those twenty-two colonies were distributed equally over all the apairies. That same season we tried out thirty colonies headed by queens of Swiss origin: they produced an average of 22 lb. against a general average of 145 lb.

It is now and again assumed that differences in colony yields are normally small and of no real significance. This may well be so where all the queens are of one strain and identical origin. But as I have all along endeavoured to stress, we can only secure positive comparisons where a considerable number of races, strains and crosses are tested side by side in identical circumstances. The larger the number of colonies, the more reliable the comparisons and the evaluations. Tests on a small scale are often grossly misleading. It will also be readily appreciated that when a selection is made embracing a relatively small number of colonies, such a breeder queen will in all likelihood prove inferior compared to one chosen from, say, ten times that number. As indicated at the outset, the bee-keeper is dealing all along with relative results and values.

I have here to revert once more to an all-important consideration: the capacity of the brood chamber and its bearing on the evaluation of performance. As we have seen already, a brood chamber that restricts the fecundity of a queen will in a corresponding measure restrict the potential honey-gathering ability of a colony. The effective strength is reduced by the restrictions to a uniform level. Minor fluctuations in performance will nevertheless arise, due to apiary accidents such as a loss of a queen, swarming, etc.; to a lesser degree due to differences in longevity, industry, wing-power or some other hereditary dispositions. However, in the essential factor, in the fecundity — which determines the potential honey-gathering ability of a colony — a levelling out or 'Gleichschaltung' has inevitably taken place. By a restriction of the breeding space not only is the possibility of a potential maximum honey-gathering ability excluded in advance, but also any chance of an exceptional performance and thereby the objective basis of a true evaluation. The assessment in such cases will be made on colonies of medium or mediocre strength, whose potential abilities remain a matter of conjecture. Evaluations made on such a basis will inevitably lead to disappointments and failures. Drifting, where no precautions are taken towards its prevention, is another factor, as I have already pointed out, that is liable to lead to completely erroneous evaluations.

It has been my constant endeavour, so far as practically feasible, to exclude every element of uncertainty and risk in beekeeping and in no case more so than in the evaluation and selection of queens destined for use as breeders. Indeed, unless the choice is based on completely reliable tests, on concrete comparisons and positive evaluations, the breeding of bees is just a blind-man's game, and no real progress in performance is likely to be achieved. Admittedly, progress may be made in the development of certain visible characteristics such as uniformity in colour, good temper, etc. but the all-important objective — exceptional performance — will necessarily elude us.

3. THE REARING OF QUEENS

a. Choice of breeders

Although I have placed much emphasis on the importance of exceptional performance, this alone cannot be the exclusive consideration in the choice of breeder queens. Exceptional performance may be accompanied by bad temper and other undesirable traits that have no bearing on performance. The queen of a bad-tempered colony will obviously not come into consideration as a breeder — unless linked with a series of outstandingly desirable traits that would counterbalance bad temper, for bad temper is a defect which can easily be bred out. However, it is clearly impossible to draw up infallible guide-lines for selecting breeders; each case must be judged independently. An intimate knowledge of the characteristics of the particular race or strain is an indispensable prerequisite. In matters of this kind much value is often laid on 'intuition', but this is to my mind a rather questionable faculty and one I should not wish to rely on myself.

I have no hesitation in stating that I know of no means or indications which would enable us in advance to determine the breeding value of a series of queens possessing an identical record of performance and equally desirable characteristics. Here again only comparative tests — in this instance between a number of sister queens, and in turn their daughter queens — can reliably and finally determine the issue. Not until the conclusion of such initial tentative tests will a breeder queen be used on an extensive scale. As a case in point, in 1973 we subjected no less than twenty-two potential breeders to such preliminary tests. Experience has shown that, among a number of sister breeder queens, there is invariably one whose progeny, whether in performance or in some other desirable characteristic, excels all the others. I refer again to the example of 1949, when twenty-two colonies with queens of a particular breeder averaged 40 lb. above an overall surplus of 145 lb. per colony. Comparisons between the progeny of different breeders provide not only much valuable information, but also give us the assurance that we are on the right track in breeding. There are, of course, breeders which fail in some way or another, in spite of every precaution taken in their initial selection, but these are found out in the tentative comparative breeding tests and eliminated before they can cause any harm of a more permanent nature, which they would do if we relied merely on 'intuition' and limited our breeding to those queens only which *appeared* to be the best. This temptation is, indeed, ever present when it is a question of matching best with the best. But who can tell which are the best breeders without the type of progeny tests I have indicated? Indeed, without such comparative tests and evaluations, carried out on the broadest possible basis, we are relying on chance and luck in the breeding of bees.

I have up to now outlined the tests on which the selection of the maternal breeders is based. The same principles hold good in regard to the paternal breeders – or the breeders that will supply the drones for the mating of the queens. In the selection and control of the male the bee breeder labours under a great handicap. We cannot tell with any certainty which colony will produce drones we desire with the particular qualities in their highest concentration.

Broadly speaking, the same principles hold good in the selection of the breeders destined to supply the drones as in the case of those used for the raising of queens. However, we clearly cannot subject the queens or the colonies meant to supply the drones to the same direct tests and evaluations. In actual fact, the evaluation of the drone colonies is limited and determined exclusively by the pedigree of the queen heading such a colony. In other words, the hereditary value of the drone is revealed in the characteristics manifested by the worker bees of the drone's grandparents. The characteristics a drone transmits are derived solely from his mother; these in turn can be ascertained from the qualities and performance of her sisters, or the worker population forming the colony from which the mother of the drones is descended. The performance of the drone colony, composed of half-sisters of the drones, cannot be used as direct evidence in judging the breeding value of the drones. However, the half-sisters should manifest in an outstanding measure, as a guarantee of their breeding value, the particular qualities we expect them to transmit. This is the final criterion applied by us in the selection of the colonies that will supply the drones.

I should mention that the recent improvements in instrumental insemination have made it possible to carry out tentative tests and evaluations with any particular set of drones, which we have found an invaluable help. This enables us to determine in advance the results of any particular mating – denoting another important step in the advancement in the breeding of the honeybee.

b. Care of breeders

So far I have dealt with the theoretical and practical aspects on which our efforts in improving the honeybee are based. I can now turn to the question of the raising of queens and the main considerations which govern our endeavours in this vitally important branch of bee-keeping.

A lifetime's experience has left me in no doubt that any interference and lack of care during the period of development from the egg until the queen is mated and attained full maturity will inevitably have an injurious bearing on her potential performance, vitality and longevity. Any serious injury will manifest itself in a sudden failure or premature supersedure; less serious damage may merely impair her laying abilities. However, a reduction in the laying powers will in reality often prove a greater drawback from the economic point of view than a premature loss of queens.

I have found that a queen which emerges in an incubator is never as good as one which spends her first few hours in her normal environment – free in the midst of a colony, even if only a small one. The difference is perhaps not obviously apparent, but it is there. So also a queen which has been caged for any length of time is seldom, if ever, as good as one which has never been confined – the extent of the injury depends on the age and condition of the queen when she is caged. Clearly any artificial devices in queen-rearing are open to objection and should be avoided.

I have dealt at length on the care we exercise in the choice of breeder queens. The next task is to ensure that a breeder is in the best possible physical condition to provide at the appropriate moment the eggs required for the raising of queens, and that these eggs are endowed with the highest measure of vitality. To ensure this the breeder queens are kept in small colonies, occupying no more than three or four Dadant combs. Their laying powers are thus restricted, which ensures the eggs they produce are endowed with the stamina needed and also prevents the premature exhaustion of the breeders. I have now and again made use of a breeder heading a fully established colony, but found without exception that the queens raised from such eggs were never as good as from those whose laying had been restricted. Apparently a queen that is producing perhaps 2,000 eggs within twenty-four hours does not possess the vitality as one which lays only a few hundred each day. Indeed, it could not well be otherwise. The same applies to queens which are being superseded – a sure indication that their life's strength is at an end. As a matter of fact, I have never come across a supersedure queen whose performance equalled those raised from eggs derived from a breeder in her prime and whose laying abilities have been restricted in the way I have indicated. I am aware this sounds all very unorthodox and contrary to the commonly held views, but our comparative tests leave no doubt on this point. Indeed, for many years now we have replaced any supersedure queens found in the honey-producing colonies in the spring.

In the breeding of domestic stock and plants of economic value the utmost care is taken to ensure that the parents are in every case in the peak of condition. No one would consider breeding from an animal or plant suffering from a debility or defect of one kind or another. Bee-keepers, on the other hand, have paid hitherto little or no regard to the condition of the queens used as breeders. Indeed, supersedure queens are almost universally considered as superior notwithstanding the fact that they are the offspring of a mother who has been in a failing condition. The nurture of supersedure queens cannot admittedly be surpassed in regard to quantity and quality, but this cannot offset the inherent deficiency in the egg, which will inevitably come to light when such queens are put to objective comparative tests. This holds equally good in the case of queens during the time they are under stress, laying to their maximum capacity, and additionally substantiated by the fact that the progeny of the superlative prolific queens are almost invariably short-lived.

Up to a few years ago the view was also widely held that the best of queens can only be raised from eggs, that is, instead of larvae. I also subscribed to this view many years ago and carried out many experiments with the object of getting breeders to lay into the artificial cell-cups. However, practical experience over the years indicated that equally good queens can be raised from larvae – provided they are no more than 18 hours old at the time of grafting. A series of experiments recently carried out at Erlangen, Germany, confirmed our own findings. While the larvae can even be a little older than 18 hours, we time them to be no more than 12 hours when the grafting is performed.

The procedure and timing to secure larvae of the age indicated is as follows: about six days before we intend to graft, the colonies with the breeder queens are fed about 2 quarts each of diluted honey. Two days, subsequently, a new but defective comb, which has been previously warmed, is inserted in the middle of each colony. Every 12 hours these combs are checked for eggs and as soon as the needed number are found preparations are made for the grafting three days later.

Before the empty combs are inserted we strive to arrange that little unsealed brood is present in these colonies so that the small larvae, destined to develop into queens, are from the moment of hatching provided with a super-abundance of food. We likewise make sure that these colonies are in a really prosperous condition and in possession of an abundance of nurse bees. Stimulative feeding is essential, unless there is a honey flow at the time. However, if the diluted honey has not all been taken down when the empty comb is inserted, it should be taken away as otherwise it will be stored in the empty comb, preventing the queen from using it immediately.

While these colonies must be kept in a prosperous condition, we do not allow them to build up to more than four combs before the conclusion of the queen-rearing season. To keep them down to the stipulated strength bees and combs of brood are taken away periodically.

I repeat: we consider no trouble too great in regard to any matter having a bearing on the quality of our queens. No effort is shirked in ensuring the most favourable conditions at every stage from the egg to the development of the fully mature queen, for it is only via the queen we possess a direct influence on the destiny of a colony and the economic advantages bee-keeping can offer. Queens that fail to come up to our expectations, that prove deficient in fecundity or vitality, in short that fail in one way or another, will invariably cause endless trouble and more often than not fail in their performance. It is a great mistake to take liberties and short-cuts in the raising of queens. Seemingly insignificant details can permanently impair the quality of a whole batch of queens.

c. Methods of cell building

There are a variety of way of getting queen cells built in small or large numbers. The method in most common use is the one by which queens are raised in a queen-right colony under the supersedure impulse. I believe this method is employed by most of the large professional queen-rearing establishments, calling for a regular supply of queen cells throughout the season. It was also used by us at one time, but not any longer now. To save time and labour and to fit in with our seasonal scheme of management we need only one or two large batches of queen cells, annually, comprising up to 600 at a time – not small regular supplies throughout the season. We require, in addition, a method of raising queen cells that will ensure positive results and one that will effectively exclude any element of uncertainty.

Queens of the highest quality can doubtless be raised by the supersedure method. This impulse is, unfortunately, subject to a wide range of influences over which we have only a very limited

92

control. During a heavy honey flow this method will usually fail. Again, to ensure the best possible results, brood must be periodically transferred from the lower into the upper brood chamber. Queen cells will usually be constructed on these combs of brood and if one is missed — an eventuality which the most careful scrutiny cannot always avoid — a whole batch of queens will be lost. Success with this method is, indeed, subject to a large degree of chance and luck than we are prepared to accept. However, as it is a method we at one time used, and more particularly as it is eminently suitable where only a small but regular number of queen cells are called for, I will describe it in detail.

d. Supersedure method

This way of raising queen cells rests on the fact that if access to any part of the hive is denied to the queen this seems to make the bees feel as if there is something amiss with her. They will accordingly construct supersedure queen cells in the part of the hive from which she is excluded. Therefore, whenever a queen is confined by means of an excluder to the lower brood chamber, bees will raise queen cells in the brood chamber above the excluder if larvae of a suitable age are available in that part of the hive. To ensure satisfactory results the second brood chamber must, of course, be adequately covered with bees and the colony must be in a really prosperous condition. Two combs of sealed brood should be transferred every fortnight from the lower into the upper brood chamber, one on either side of the frame or comb with the queen cells. Any queen cells constructed on these combs of brood will have to be destroyed without fail, about seven or eight days subsequent to their transfer into the second brood chamber. While the building of queen cells is in progress, the colony must be constantly fed small doses of diluted honey when there is no natural flow.

The foremost advantage of this method is that any colony can be used for the building of queen cells without any special preparations, apart from those indicated, and at no loss to the honey crop. Furthermore, such colonies will build a continuous series of queen cells, at intervals of five days, provided the sealed brood is replaced periodically. However, depending on the strength of the colony, one should in no case be tempted to raise more than ten or a dozen cells at a time. The fact that this method is almost universally used by the largest queen-rearing establishments is clear proof that good results are assured when properly applied. It is liable to fail during a heavy honey flow, for then the bees will usually concentrate all their energies on the gathering of honey and will in such circumstances on occasion even neglect any queen cells under construction.

e. Our own method

There can be no doubt whatever that the swarming impulse provides the best nurtured and best developed queens, for when a colony prepares to swarm it has reached an optimum in its organic development, as well as an opulence in every direction. Indeed, swarming is the natural manifestation of a colony having reached the summit of affluence. In such circumstances ideal conditions prevail for raising the best of queens from the physical point of view.

While we cannot make ourselves dependent on the normal swarming season and swarming impulse for the supply of queen cells required, we can create conditions that will give rise to the swarming impulse and an optimum affluence whenever we so desire. The procedure is as follows: a second brood chamber, containing ten or twelve combs of brood and adhering bees, is placed above an excluder on a queen-right colony which in turn must be in possession of a minimum of ten combs of brood. The future cell-building colony thus set up possesses no less than twenty Dadant-size combs of brood. In times of dearth, when the bees are unable to collect any nectar, feeding must be resorted to immediately of no less than 1 quart of diluted honey each day.

Nine or ten days later all the queen cells constructed on the combs of brood in the second brood chamber must be destroyed without fail. To make quite sure none are overlooked part of the bees are shaken off as each comb is checked. In the lower brood chamber, to which the queen has been restricted, preparations for swarming should by now be under way and three or four days later the huge colony, teeming with nurse bees, will have attained the stage when it will be in the best possible condition for the construction of queen cells.

The final date of the grafting is fixed by the age of the eggs and larvae, which we were able to determine three days beforehand. About 9 a.m. on the day the grafting is to take place the final steps in the setting up of the cell-building colony are made. The upper brood chamber is set in place of the one with the queen, the latter is set down some few feet away from its original position. The queen is then searched for and when found the bees from seven or eight combs of brood are then shaken in front of the hive on the old site. At the conclusion of this operation the prospective cell-building colony will be in possession of the greater part of the field bees and a huge concentration of nurse bees, together with an abundance of stores. The gigantic colony will be in a highly developed state of the swarming fever and, in conjuction with the removal of the queen and all unsealed brood, in an ideal condition for attending to the grafted larvae. A colony thus set up will take to the grafted larvae immediately. There is no likelihood of their being neglected during the first few hours, as is liable to happen in other cases. A frame with three carriers of twenty grafted cups each is given within an hour or two subsequent to the final setting up of the colony. In the absence of a honey flow, feeding must be continued at the rate of about 2 quarts each day until the queen cells are sealed on the sixth day after grafting. This is absolute imperative. Should there be a flow the colony may need additional room. However, we find the best results are secured, regarding the cell-building, when the bees feel crowded.

The hive with the queen is taken to another apiary and fed a gallon of syrup immediately. If thus treated, it will suffer no diminution in strength by the time it is returned to its original site a week or eleven days later.

We usually have twelve of these cell-building colonies in operation simultaneously at the end of May or early in June. As soon as the queen cells are sealed, by the afternoon of the sixth day after grafting, the sealed cells are transferred into three of the colonies; each will have three frames holding four cell-carriers respectively, or approximately 200 queen cells. The nine colonies, from which the queen cells were removed, are next morning taken to the apiaries from where the bees and brood originally came and are now returned to their respective hives or any one in need of help. The queen-right colonies, which were taken away on the morning the cell-building colonies were set up, are simultaneously returned to their permanent site.

As already indicated, our queen-rearing is concentrated into as few operations as is possible. They are, moreover, carried out to a pre-determined timetable and independent of the weather and possible honey flows. In an undertaking of this kind, on which the ultimate success of our bee-keeping is based, every uncertainty and element of chance and luck must necessarily be excluded as far as possible. We have found that the method I have just outlined fully meets all these requirements and will produce the best possible queens from the physical point of view.

f. Number of queen cells per colony

The severely practical bee-keeper is of course less concerned regarding the actual number of queen cells he can secure than in the ultimate physical quality of the queens. Poorly-reared queens are the bane of modern bee-keeping. Acceptance of the grafted cups is largely determined by the strength, composition and condition of a cell-building colony and by racial and hereditary dispositions. Hybrids generally do better than pure-bred stock. The Western European varieties, also the Carniolan, will build relatively few cells. On the other hand, colonies of Oriental origin, particularly the Armenian variety, will raise 200 to 300 queen cells at a time, seemingly without the slightest diminution in the quality of the resulting queens.

Objectively considered, size by itself does not give us any indication of the fecundity and vitality of a queen. Those of Western European descent, also again the Carniolan, usually surpass in size the queens of most other races. Yet, when compared to the others, they are the least prolific – which most certainly holds good in regard to our former native bee. Cyprian queens are small, but nevertheless extremely prolific. However, a Carniolan queen, no larger than a Cyprian, would in all likelihood prove unsatisfactory.

Our queen-rearing operations are based on a pre-determined minimum acceptance. This is essential, as otherwise we would not be able to plan the various operations to a fixed timetable. In the case of the cells built under the supersedure impulse in a queenright colony, occupying two brood chambers of BS size, we based our calculations on a minimum acceptance of nine cells from a batch of fourteen cups; the actual average being about eleven. In the case of the second method, the queenless colony composed of a huge force of nurse bees, we expect a minimum acceptance of forty-five queens cells from sixty grafted cups; the actual acceptance averages approximately fifty-five cells. We aim to have a substantial surplus of queen cells for disposal to meet any contingency, without overtaxing the potential nursing ability of the cell-building colonies. The queenright cell-building colony will accept a batch of grafted cups every five days, whereas the queenless colony will prove only satisfactory for a single batch. A second batch could be raised, but the queens would not come up to the high standard we demand. The actual number of high quality queen cells a colony is able to raise is subject to a combination of factors: foremost on the actual number of nurse bees available; on their hereditary and physiological dispositions; and no less so on the degree of affluence prevailing.

In this context I must re-emphasise the overriding importance of stimulative feeding in the absence of a honey flow. A light honey flow will give by far the best results; in its absence the feeding of diluted honey, notwithstanding the cost, is called for. The dilution should not exceed one-third of water to two-thirds of honey and, to prevent any fermentation setting in, no more than a gallon should be fed at a time.

4. CONTROLLED MATING

As in all spheres of selective breeding, we cannot hope to achieve any positive improvement in the honeybee in the absence of an effective control of the maternal as well as the paternal progenitors. We can improve our stock by the introduction of a queen of known performance. Indeed, by merely requeening our colonies with queens raised from a breeder of a highly productive strain we can substantially improve the standard of our stock notwithstanding the random matings. Furthermore, if such a course is pursued over a period of years, the drones flying from those colonies will in time raise the general standard of the stock in the neighbourhood. But we are thus merely putting to good use a strain of known performance. We are not improving the honeybee itself; we are not doing any breeding in the strict sense of the word; our efforts merely entail a propagation of superior stock – a most commendable endeavour and the only course open to the vast majority of bee-keepers.

There can clearly be no breeding worthy of the name, interpreted in the strictest sense of the word, except where complete control over both parents is at our command. This obviously holds equally good in the case of the honeybee, notwithstanding the well-nigh insuperable difficulties. However, our experience seems to indicate that something of value can be achieved by a limited control of the drones. More exacting endeavours demand mating in isolation, or instrumental insemination where absolute individual control is called for. The first possibility indicated offers some scope to anyone operating a number of apiaries.

Before the establishment of our mating station we secured satisfactory selective matings by heading every colony at one of our out-apiaries with queens of a particular line. Forty colonies will raise on an average 50,000-80,000 drones. While this number will not necessarily ensure correct matings in every case, the overall results obtained were eminently satisfactory. This method of partial control is of particular value in instances where a professional bee-keeper wishes to secure a particular productive cross, for example, Italian queens mated to Greek or Carniolan drones. A limited drone control, as here indicated, can offer substantial economic advantages involving a minimum of outlay and effort. Anyway we found this form of control of great practical value. But we realised that only matings in complete isolation would meet our more exacting requirements fully. The close proximity of Dartmoor provided the needed facilities.

a. The mating station

We were fortunate in finding a suitable site in the heart of Dartmoor at a distance of ten miles from the Abbey, situated in a sheltered valley on the side of a hill facing south-east and surrounded by a belt of trees on three sides, protecting it most effectively from the prevailing south-west and north-westerly wind. As a mating station it is quite isolated, the nearest bees being some six miles away. The area is, furthermore, almost devoid of trees and habitations and the open moor so bleak that no stray swarm could survive for long. The altitude of 1,200 ft. above sea level materially contributes to the isolation. Established in 1925, it has been in continuous use ever since.

While we have the needed isolation, a site in the heart of Dartmoor inevitably entails many drawbacks and problems. The severe winters and general inclement climatic conditions have, on the one hand, the advantage of quickly showing up any hereditary defects or errors in breeding. On the other hand, the actual mating results are now and again most disappointing, due to the long periods of dull sunless conditions and low temperatures. Indeed, it has happened before now that an entire batch of 520 queens failed to mate due to a spell of continuous unfavourable weather extending over a period of six weeks. Even in an average summer we cannot hope to secure more than two consecutive batches of mated queens. But these drawbacks must be set against the overall advantages secured. The outstanding successes achieved could not possibly have been attained in any other way. As in every field of endeavour, nothing of real value can be accomplished without a corresponding effort and outlay.

Needless to say, the running of an extensive mating station demands equipment of special design and methods of management. When we set up our mating station we were virtually entering an unexplored region of bee-keeping. We had no reliable information to guide us. In fact, the little information then available, on how to operate a mating station, proved in the main grossly misleading. The great movement – die Rassenzucht – inaugurated by Dr U. Kramer in 1898 in Switzerland was, as is so often the case, carried away by enthusiasm and ideals incompatible with stern realities. We were therefore compelled to tread the thorny path of many disappointments and failures before success came our way. At that time little was known on in-breeding, on cross-breeding, on the flight range of drones and the number of drone colonies required, on a safe way of transporting queen cells, on the most suitable type and size of nuclei hives, and a host of other technical problems.

b. Nuclei hives

One of the foremost problems confronting us was the type and size of hive. It had to be small enough to prevent the raising of drones, but at the same time of a capacity to render the nuclei self-supporting so that they should need as little attention as possible. We had already some experience with miniature baby nuclei of the type widely advocated at the turn of the century in North America and on the Continent. We were therefore able to avoid this particular pitfall. We knew we could not hope to secure queens of the highest quality in miniature colonies composed of no more than a few hundred bees. So we proceeded to test a variety of hives, holding a single and alternatively a number of nuclei; also nuclei on one comb and on a series of combs. These trials extended over many years before we were in a position to make a final choice.

For about twelve years we used a hive which accommodated four nuclei side by side, each on three half-size BS combs. This particular pattern served our purpose admirably. In fact we still have thirty of these hives in use today. They were, however, too small for wintering, except when united. But in the course of the years I gradually came to the conclusion that a way had to be found which would enable us to winter our annual production of queens so as to enable us to do our requeening in spring. We finally decided on a hive holding sixteen half-size Dadant frames, divided into four compartments holding four combs each. The changeover was made in 1937 and the experience and results gained since then have demonstrated that this was the right choice.

We have a hundred of these hives in use as well as thirty of the original type, holding half-size BS combs, making a total of 520 nuclei when the mating station is in full operation.

c. Transporting queen cells

At every step in our bee-keeping we have endeavoured to eliminate any avoidable element of risk. This holds particularly good in regard to the handling and transportation of the queen cells until they are safely placed in the nuclei. Immature queens are easily injured when they have to be transported a distance of many miles and no less so by an undue exposure to the sun and low temperatures. As we have found, one cannot take too much care for the injury is in every case of a permanent nature.

We leave the queen cells in the cell-building colonies until the queens are due to emerge within twelve to twenty-four hours. On the eleventh day after grafting – in very warm weather, on the tenth – the whole cell-building colony with the queen cells is transported to the mating station. On arrival one cell-carrier, with fifteen to twenty queen cells, is removed at a time and placed into a basket fitted with a warm water tank, on top of which the cells are wrapped in a blanket to keep them warm while the distribution to the nuclei is in progress.

To ensure that the young queens are in no way injured en route to the mating station, the hives with the queen cells are placed on a 3 in. deep foam rubber mat to absorb any jolts and vibration. The tyre pressures are in addition kept as low as at all possible – all very needful and appropriate precautions.

The bees comprising the cell-building colony are used for strengthening the nuclei or drone colonies. They have therefore to be passed some time beforehand through a queen excluder to eliminate the unwanted drones. This is generally done on the afternoon of the sixth day after grafting, before the queen cells are brought together as already described. This is accomplished in the following way: the brood chamber with the bees is set aside and an empty one put in its place with a zinc excluder between the bottom board and empty brood chamber. The bees are then brushed off the combs in front of the entrance and allowed to re-enter through the excluder. Five of the combs are returned to the empty brood chamber, along with four foundations of which two are placed on either side. This allows room for the three frames with the cell-carriers, each of which is placed between two combs of stores.

d. Drone colonies

Up to quite recently only one drone colony was allotted to each mating station on the Continent with the object of getting the best queens mated to the best drones. It is, however, now recognised that such a course led to untold failures. Suspecting this would be the case, we at no time provided fewer than four drone colonies at our mating station, apart from a number of special experimental matings.

The queens heading these colonies are in every case sisters, chosen from a large number. Though of identical parentage there will nevertheless be some variation in the drones produced. This will in turn allow a wider selection in the offspring resulting from these matings, modifying in some measure the risks of in-breeding and extreme uniformity.

We do not necessarily keep the same set of drone colonies at the mating station throughout the season. When found desirable they are exchanged with another set possessing drones of a different race or strain. In very special cases the drones are individually selected before they are taken to the mating station. This latter method is an exceptional procedure and now virtually superseded by instrumental insemination.

A word of caution in regard to the number of drones a colony can raise and nurture adequately must here be added. I believe it is not commonly realised that the raising and upkeep of a large drone population will tax a colony's nursing and physical resources severely. We restrict the production of drones to two-thirds of one comb per colony, apart from the drone comb found normally in colonies. We furthermore take the utmost care to ensure the drone colonies are at all times in a state of maximum prosperity and affluence. The utmost vigilance is called for to maintain them in the peak of condition, particularly in the case of inbred strains. In prolonged periods of dearth, feeding is essential.

e. Instrumental insemination

While instrumental insemination will in all likelihood never replace the need for mating stations, the recent improvements made in the technique have placed at our disposal additional facilities of control of inestimable value. Tentative tests, which were hitherto impossible, can now be freely carried out with absolute reliability and certainty.

ADDITIONAL CONSIDERATIONS

In retrospect there seems a need to draw attention to a few special aspects of our beekeeping that we consider of far-reaching importance, but we fear their practical significance is often overlooked or not fully appreciated. Our method of management, the special hive and equipment we use, would be of little practical value without queens of the highest physical quality and a corresponding genetic endowment, for apart from the local nectar sources and varying climatic factors the potential honey gathering ability of a colony is largely determined by the fecundity of the queen, or more precisely by the actual colony strength. Whilst most amateurs feel pleased as long as they secure some surplus from their colonies, the more progressive beekeepers and more particularly those who derive their livelihood from the keeping of bees will strive to secure maximum returns, in conjunction with a minimum expenditure of effort and outlay from each colony. We all know that the most productive colonies invariably demand the least care and attention. We regard an adequate genetic fecundity as the linchpin of our beekeeping.

As experience has shown, a second rate queen or one of mongrel origin may appear satisfactory in a hive of limited capacity, but would lamentably fail in a hive of the size used by us. This failure would show up still more glaringly in the respective yields of honey secured. Indeed had we not made the change to the Modified Dadant hive in 1924 we would in all likelihood have remained in blissful ignorance of the vital importance of fecundity and the many other discoveries of practical value. Clearly wherever the genetically determined fecundity is prevented from manifesting itself fully the basic criterion for a potential honey gathering ability is missing and a reliable selection for a progressive breeding scheme absent. The realization of this fact has played a major role in our breeding endeavours over the years.

Among the discoveries made, we found that the genetically determined fecundity of a queen is subject to many adverse influences during the period of development from the egg until a queen reaches full maturity some weeks after she has started to lay. Any injury suffered in this period is permanent, and if severe will lead to premature failure and supersedure, causing endless extra work and consequential losses.

It is very strange that beekeepers tend to ignore certain universally acepted axioms that are regarded as inviolable in all similar spheres of endeavour. In the breeding of livestock no-one would ever consider using an animal in failing health, or, in the production of seed, employ plants obviously defective. However in beekeeping it is widely assumed that supersedure queens are superior to any others, disregarding the fact that in every instance of this kind the mother is suffering from some major weakness or other. Positive comparative tests carried out by us on the widest possible scale have conclusively demonstrated the erroneousness of such an assumption. We have in fact never come across a supersedure queen to match the performance of a queen of our own raising. We have accordingly for years replaced every supersedure queen when our annual requeening is carried out in spring: their unclipped wings forming a positive mark of identification.

I should perhaps stress that the findings here put forward are in every case the result of tests carried out over a period of close on seventy years. Moreover these tests embrace every known race and cross of the honeybee. They were also made in a hive which imposed no restrictions on the fecundity of the most prolific of queens, an essential condition to comparative tests of this type. Every precaution was likewise applied to prevent any drifting, which would place the results secured in doubt. Any tests lacking these essential provisions and safeguards would prove valueless and misleading.

The actual maximum potential fecundity is in every case a racial characteristic. It is however one that is subject to a variety of adverse influences and circumstances. On the other hand, it would be wrong to conclude that the most robust of queens would prove the most fecund. Indeed, the queens of the most fecund races, such as the Cyprian and Syrian, are relatively small in size. This is likewise a genetic trait of our own strain. As practical experience has shown, extreme fecundity is usually associated with a number of major drawbacks. Bees of the super-fecund bright yellow strains are usually short-lived, highly susceptible to acarine and suffer from a loss of stamina. It may be safely assumed that the loss of stamina and longevity is a direct result of the super-fecundity.

According to our observations and findings, queens capable of laying a maximum of about 3,000 eggs per day at the height of the season appear to give the most satisfactory results. However it stands to reason that eggs produced in such numbers cannot embody the vitality and stamina as when a queen is restricted to no more than 300 a day. Here again our findings fully substantiated this assumption. Indeed we had never had much luck when larvae for grafting were taken from a colony where the queen's laying capacity was extended to her maximum limit. We therefore keep all breeder queens in nuclei of no more than four M.D. combs during the period the grafting is done. Also, to prevent any overcrowding, combs of brood and bees are taken away whenever deemed necessary. To secure top quality queens, the restriction of the breeders as here indicated is to our mind essential.

Colonies with queens more than two years of age, usually manifest a progressive flagging and a general deterioration in zest and stamina. In cases of supersedure we know with certainty that a queen is failing. A deterioration due to old age is usually more difficult to determine, particularly when a breeder is restricted to a few combs. Even one in her 5th year may in such circumstances fail to manifest any visible deterioration. Clearly old age is a loophole by which queens of inferior quality may find entry, if the necessary safeguards against such a possibility are disregarded. Whenever in doubt we would give a breeder of lesser age and proven vigour the preference. Longevity is admittedly a trait of great practical importance, but this trait will have been transmitted unimpaired to the daughters raised whilst the breeder in question was still in her prime. The positive proof of old age we would regard as a confirmation of longevity and we would record it accordingly.

When grafting, the actual age of the larvae is another factor with a bearing on the quality of the queen raised. Here again our experience seems to indicate that to ensure queens of the highest quality the larvae should not be more than twelve hours old. The acceptance of the young larvae will not be quite as good as when older larvae are grafted, but quality is at every stage the more important consideration. The actual number of top quality queen-cells which a colony is able to raise is primarily determined by the force of nurse bees available and the general affluence of the colony. The particular race of bees used also plays an important role. Needless to say, nothing must be left to chance in the preparation of a cell-building colony, for any false economy in this connection would render all other precautionary measures and efforts useless and lead to endless difficulties and disappointments subsequently. The temptation to consider any queen-cell as 'good enough' is obviously one to which many amateurs succumb all too readily.

We have so far considered the factors that have an influence on the physical quality of a queen from the moment the egg from which she derives her existence is laid, until the young queen emerges from her cell, in other words, the initial period of her devlopment. However, in the subsequent period, until she reaches full maturity some weeks after she has started laying, every queen is subject to many risks and adverse influences. At this stage the harmful influences and dangers are mostly brought about by the beekeeper, ostensibly in the name of progress or an endeavour to secure some advantage or other. The use of miniature mating boxes is one of them; the exclusive use of candy, necessitated by mating boxes of this kind, is another. The emergence and confinement of the young queens in cages is so likewise. As practical experience has demonstrated, to secure queens of the best quality by such means is an impossibility, except perhaps in ideal climatic circumstances.

Cageing the young queens within a few days of their starting to lay, for the purpose of introducing them to other colonies or to despatch them by post, will inevitably harm a young queen to a greater or lesser extent. On the other hand, if they are first allowed to lay undisturbed for some weeks, the risk will be minimal. But if a queen is caged for any length of time she cannot escape some injury. She will never prove quite as prolific as if she had never had suffered a lengthy caging. The actual introduction, unless carried out with the greatest possible care, usually entails the loss of many queens at the very threshold of their useful existence. Of those accepted far too many are injured in one way or another. The immediate and consequential losses thus sustained must be regarded as one of the most tragic aspects of modern beekeeping. However, with adequate care, the safe acceptance of queens poses no real problem. The system we adopted, as outlined in the main section of this book, does not in reality entail any introduction in the commonly accepted sense of the term, but merely an exchange without any loss or injury to the queens.

The precautionary measures thus indicated, to ensure queens of the highest quality, may appear to many 'counsels of perfection'. In reality they are no more than elementary common sense measures which moreover do not entail any special effort or expenditure. On the other hand they ensure, if observed, queens of the highest quality: queens that will render beekeeping more pleasureable and remunerative. All too often beekeepers place a great emphasis on matters of little or no importance and the things that really matter are overlooked.

The concept of 'top quality' in this connection denotes not only a high physical perfection in the queens but also a commensurate genetical endowment. Mongrels usually possess an exceptional constitution and an ability to eke out an existence in the most adverse seasons without assistance of any kind, but fail lamentably in most other respects. Modern beekeeping demands a prolific, good tempered, non-swarming bee, capable of making the most of every honey flow. As experience has demonstrated the actual superiority of stock of this kind, as also that of a good first-cross, given the appropriate management and care, can be startling to the point of incredibility. However only when the necessary conditions are provided that ensure the full unimpaired development of all the genetic faculties, especially fecundity, are maximum returns per colony possible.

In every sphere of endeavour half measures and haphazard ways will inevitably lead to endless difficulties and disappointments. Breeding and raising top quality queens offers no exception, but seemingly insignificant common sense precautions can make vast differences in the yields of honey per colony and render the keeping of bees ever more attractive and rewarding.

CONCLUSION

I have in the foregoing pages tried to outline as briefly as possible the main points of our bee-keeping: the particular hive we use; the seasonal management adopted by us; the strain of bees we have evolved; and the methods of breeding applied to ensure queens of the highest ability of performance. I have also endeavoured to indicate the reasons why one course of action or one particular alternative was adopted in preference to another. I have likewise shown that we strive to interfere as little as possible in the activities and organisation of a colony, and that the honeybee will at all times blindly follow her instincts regardless of our wishes. Indeed, the old-time title 'beemaster' has no real validity in modern beekeeping. The tasks of the modern bee-keeper might more aptly be described as a 'service'; in fact, we are more truly servants than masters.

Our success at Buckfast has not been achieved by the use of complicated devices or involved methods of management. On the contrary, every item of equipment not really necessary, as well as all needless manipulations, have been rigorously eliminated. The emphasis has been on essentials in every case.

The constant emphasis on the economic aspects of bee-keeping may possibly have conveyed the impression that at Buckfast no value is placed on the aesthetic aspects of bee-keeping. But anyone familiar with our endeavours will know that this is not so. I personally find the daily contact with bees, observing the unfolding of each colony's individuality, the differences between one race and another, the untold nuances manifested in their various traits and manner of behaviour, by far the most absorbing and fascinating aspect of bee-keeping. In any case, the real interest in handling and keeping bees is not dependent on an unwarranted expenditure and futile efforts, but rather on a true understanding and appreciation of the ways and needs of the honeybee. A wise use of the facilities at our disposal is in reality the basis on which success in bee-keeping depends. True idealism and economic interests do not conflict but are complementary and success is the most compelling force in any human enterprise. The keeping of bees forms no exception.

MEAD *

One of the oldest beverages in the world is mead. Long ago in the simple and more thriftful days the economy of every home was self-contained. Foreign foods and foreign beverages played little part in the life of ordinary people. Bees were kept, not only for the honey they provided, but for the wax that gave light on winter nights. Honey was primarily a sweetmeat, or sweetening agent, and the only one available before sugar could be obtained freely. Waste is a modern characteristic! Every bee-keeper knows that much honey is still wasted in extraction. Our forebears washed the pieces of comb, after all honey possible had been retrieved by crushing and draining, and from the washings they brewed a beverage. It was so highly esteemed that wine was regarded as less delicate and delicious.

On many a winter night, with a wax candle on the table and a fire of logs on the earthen hearth, the old bee-keeper entertained his neighbour or solaced his own soul by a cup of mead. The home was simple, but the fare was rich; the talk around the fire most companionable, and the liquor most stimulating.

For many a year there stood on Salisbury Plain a shepherd's hut on wheels, among bushes of gorse, and surrounded by hives. Few passers-by would notice the hives among the gorse, and possibly not even see the hut, unless the gale blew ribbons of smoke across the traveller's path and drew his attention to it. In that hut lived a solitary man who had travelled far, but had found anchorage at last among his bees on Salisbury Plain. He sold his honey, and made his candles, and brewed his mead. At night, let the storm howl as it might, though the hut shook and trembled, he

* First published in BEE WORLD 34(8) : pp. 149-156 (1953).

was content to lie on his bunk and read the pages he had written, and sip the mead he had drawn from his cask in the tiny shed outside. It was a work on philosophy of possibly little value, but the man was in essence a philosopher. The mead warmed his lonely heart, and was the only concession to a luxury he otherwise despised. He let the world go by, and he coveted none of the pomps and delicacies that others spent so much labour to acquire.

In every cottage in England similar contentment centred around the mead that came so readily to the hand of the beekeeper and, when all were bee-keepers, it was the equivalent of the sherry, port, or brandy of richer folk in a later age. But it now has become quite the fashion to serve mead on even wealthy tables again — the old-time value is being recognised afresh.

We have made mead ever since we kept hives, because it is one way of using the honey which would otherwise be thrown to waste. It is necessary to follow strictly certain rules in the making of mead if the best results are to be obtained. Before detailing our findings, may we say that we have produced four distinct varieties of mead – one that is of the ordinary type of table wine or light sherry; another of a dessert type, richer and sweeter than the first; a spiced wine that possesses a character all its own; and, lastly, a sparkling wine similar to a sweet champagne. Each has its admirers, and honey and water are the sole ingredients.

Special points in making mead

Before describing the actual mode of preparing the 'must', and the sterilisation, fermentation and maturing of mead, I will first, for the purpose of a clearer understanding of the whole process, briefly set forth the points which according to our experience are of the utmost importance in the making of mead.

(1) It is most essential to use soft water – clean pure rain water or, alternatively, distilled water. Water from the tap, even if soft, is not suitable on account of the particles of rust it may contain.

(2) The kind of honey used determines the flavour and bouquet of the mead. Mild-flavoured honeys, such as clover or

lime, are the most suitable. A strong-flavoured honey – excepting heather – should never be used; it will merely lead to disappointment. There is a view prevalent that any strong-flavoured or off-grade honey which cannot be used for any other purpose is suitable for making mead. This is, however, a completely mis taken notion, and it is the cause of many failures. Only the very best mild-flavoured honey will produce the finest vintage mead.

(3) The yeast used largely determines the character of a mead. A pure-culture grape wine yeast gives the best results. There are a number of pure-culture wine yeasts which give particularly good results in the fermentation of mead. We use a Maury wine yeast. Hitherto, brewer's yeast or baker's yeast has been almost universally recommended as a ferment for mead. The former imparts to the mead, as would be expected, a distinct beery taste; the latter gives the mead a tang peculiar to this type of ferment.

(4) The solution of honey and water must be sterilised, and also the vessels or cask in which the process of fermentation is completed. Moreover, all possibility of any contamination subsequent to the sterilisation must be carefully avoided. This is really of the utmost importance.

(5) Chemicals or special nutrient salts, as generally recommended, should be used with the utmost circumspection. Such additions will greatly hasten fermentation, but tend to spoil the delicate flavour and bouquet of a high quality mead.

(6) The temperature throughout the process of fermentation must be kept constant, neither too high nor too low. The most favourable temperature to the species of ferment I have recommended, the Maury wine yeast, is 65°-70°F.

(7) The process of fermentation should be effected during the summer months, for at that time the temperature and atmospheric conditions are most favourable to the propagation and growth of the yeast cells. Therefore, the best time to begin operations is from May to July. This applies to all still wines. In the case of sparkling mead, the most suitable time to start fermentation is in October.

(8) To obtain a mead of character, a beverage that will surpass the finest wines produced from the juice of the grape, it must be matured in sound casks made of oak, and it must be stored in

wood a minimum of five years before it is bottled. This is particularly important in the case of a sweet mead.

Proportion of honey to water

It should be clearly understood that the flavour, bouquet and character of a mead are the combined products of the floral essences contained in the honey and the particular type of yeast used. On the other hand, the body or 'oiliness' of a mead and its alcoholic content are largely determined by the proportion of honey to water used. A mead made of less than 2 lb. of honey to 1 gallon of water will not keep. The maximum concentration which can be effectively fermented is 6 lb. to 1 gallon of water. For a sparkling mead I would recommend 2¼ lb.; for a dry, still wine 2-3½ lb.; for a medium-sweet to a rich dessert wine 4-5 lb.

Whenever the proportion of honey in a solution is not known, as when honey is washed from cappings, it will be necessary to use a hydrometer to determine the honey content. Alternatively, the proportion of honey may be ascertained by the less accurate method of weighing a quart of the liquor. The weight of a quart of liquor and the amount of honey per gallon of water, together with the specific gravity readings (at 60°F) and the actual sugar content are as follows:

Weight/quart	Honey/gall.	Sp. gr.	Sugar content
2 lb. 10 oz.	2 lb.	1.053	13.04%
2 lb. 10½ oz.	2½ lb.	1.064	15.61%
2 lb. 11 oz.	3 lb.	1.075	18.13%
2 lb. 11½ oz.	3½ lb.	1.086	20.60%
2 lb. 12 oz.	4 lb.	1.096	22.81%
2 lb. 13 oz.	5 lb.	1.114	26.06%
2 lb. 14 oz.	6 lb.	1.128	29.66%

We gauge the proportion of honey – or really the sugar content – for every sample by a specific gravity reading. It is the only accurate way for the purpose of comparative experiments, as the water content in honey itself varies from 17 to 25 per cent. Moreover, a solution of ling-heather honey has to be filtered to exclude all foreign substances before it can be used for making mead. If it is not filtered, the foreign matter will spoil the flavour, giving the mead an unpleasant tang. So the mixture of heather honey and water is first sterilised (only just brought to the boil), then after it has cooled it is strained through a filter bag. We use one made of raised milk filter cloth, but filtering material made of nylon will serve the purpose equally well. Muslin and any of the common straining cloths are, however, unsuitable. The readiness of heather honey to filter varies from year to year, and at best it is a slow process. Two gallons of the solution, the capacity of our filter bags, usually take about 24 hours to filter. If at first some sediment should pass through the filter, which is usually the case, then this is returned and re-filtered. When well filtered the liquor obtained should be crystal-clear and of a rich deep golden hue. After the filtering is completed it is necessary to re-sterilise the must. It is advisable to make the initial solution in a concentrated form, and then dilute the filtrate to the desired density before the final sterilisation. A honey-must thus prepared will develop, with good vinting and careful handling and keeping, into a mead of supreme quality.

I now come to the actual making of the mead. It is necessary first of all to obtain a sound cask and one that has contained wine – preferably a sherry cask. After washing it thoroughly with hot water, without the addition of any cleansing agent such as soap or soda or any ordinary kind of detergent, we can make the honey-must which is to be the raw material of the mead.

I again emphasise very strongly that pure water, soft and pure as from rain itself, together with honey of the best quality, are the absolute necessities for a mead of finest character. Honeys derived from clover, the limes or ling undoubtedly make mead of supreme quality. But even so, just as honey from the same source varies from season to season, and from district to district, so will the mead vary. Honey from old brood combs should not be used. The mead would be inferior. But honey from cappings, if of good flavour, is quite suitable. Soft warm water is poured over the cappings, and after all the honey has been dissolved, the wax is then formed into balls and well squeezed, and every drop of honey is thus saved. In whatever way they honey is obtained – from cappings or otherwise — it is well to use a hydrometer for ascertaining the exact density of the solution. The specific gravity we use for a light table wine is 1.055; if the reading is taken after the final sterilisation it should be 1.058.

Having mixed the honey and water solution, it is then sterilised. The must is brought just to the boil and held at that point for a minute or two. It should on no account be boiled for half an hour, as is generally recommended, for the prolonged boiling would dissipate the subtle floral essences contained in the honey. It is these very delicate aromatic oils that impart to mead its inimitable bouquet and aroma. The object of the sterilisation is to kill all the yeasts contained in the honey, or any which have entered from the surrounding air, which if not destroyed would spoil the mead. The sterilisation should be done in a vitreous or tinned utensil, or a container made of stainless steel or aluminium. A vessel made of copper, if perfectly clean, is also suitable. But the honey-must should never at any stage come into contact with iron or brass or galvanised metal.

Practically every recipe for the making of mead recommends the addition, at this stage, of various inorganic salts to hasten the process of fermentation. But our findings over the years have indicated that the best results are obtained not by a rapid but slow fermentation. We have also found that cream of tartar and citric acid, when used in moderation, will foster the fermentation without impairing the quality of the mead. We use 2½-5 oz. of cream of tartar and ½-1 oz. of citric acid to every ten gallons of must, depending on the quantity of honey per gallon of water.

After the completion of the sterilisation the honey-must is poured into the cask while still hot. This will destroy such yeasts as may be in the cask itself. The cask should not be filled to the top, but an allowance of about 3 in. made for the addition of the yeast culture and the subsequent expansion of the liquor when fermentation sets in. The bunghole is at once plugged with a tightly packed wad of cotton wool, to exclude all micro-organisms from the liquor. The wad of cotton wool is replaced with a fermentation valve after the insertion of the yeast.

Addition of the yeast

The liquor is allowed to cool until it has attained a temperature of 80°F; the pure-culture grape wine yeast is then added. If the wine yeast is in a liquid medium, the amount should be not less than 1 per cent of the total quantity treated. A yeast culture supplied in a test tube has, however, to be brought into active propagation before it can be added to a honey-must exceeding five gallons in quantity. This is to ensure a rapid and satisfactory onset of the fermentation, which is essential.

The fermentation

After adding the yeast culture the stormy fermentation will begin within about thirty-eight hours, but if the yeast is not in an active state, or if too small a quantity is used in relation to the bulk of the honey-must, several days may elapse. The first violent fermentative process will gradually decrease in the course of a few days. The primary fermentation will then supervene. This second process will take six to eight weeks in the case of a light mead, but in a heavy mead it may extend over a period of three to four months or even longer.

On completion of the primary fermentation, of wine made from grapes or the juice of various fruits, the clear or partially clear liquor is decanted from the lees or sediment. This is no doubt essential in all such cases, owing to the nature of the sedimentation. But I am not convinced that racking or decanting is an advantage in the making of mead. We prefer to mature our mead, particularly the sweet type, on the lees for a period of years. Sherries are not removed from their lees and it is claimed that it is this which imparts to them their particular quality.

After the fermentation the mead should be allowed to mature in wood for at least two years. To bring it to perfection an even longer period is necessary. But both light and sweet mead make a pleasant and palatable drink as soon as they have fully clarified.

It will be found that in the case of a full-bodied mead another very slow fermentation will take place the year after it is first made. The minute quantity of carbon dioxide formed by this after-fermentation normally escapes through the wood.

Three distinct stages in the process of fermentation take place:

(1) The stormy fermentation, which begins within thirty-eight hours of the insertion of the yeast culture and lasts about three days.

(2) The primary fermentation, which may take six weeks to four months or longer – depending on the vigour of the yeast and the proportion of honey contained in the liquor.

(3) The after-fermentation, which in a light mead may be hardly perceptible, but in a full-bodied one may reassert itself during the summer months for one or two successive years.

Clarification

Meads made of certain honeys seem to take longer to clarify than others, notably that made of clover honey. This honey often forms very tiny flaky particles, like those found in some Rhine wines. If any difficulty is experienced in this connection the mead may be fined with isinglass. For every 10 gallons, ¼ oz. isinglass is well beaten up in a little mead and then stirred into the cask. In a few days the mead will be perfectly clear. However, I prefer to leave the clarification entirely to time.

Vessels

I believe it is not generally realised that mead cannot be made to perfection in small quantities – in glass jars or bottles of small capacity. It certainly cannot be matured in stoneware jars or glass or in a plastic container. Like all wines and alcoholic liquors, mead can only be properly matured in wood.

Jars holding two to ten gallons are ideal for fermentation. In glass the air is completely excluded and, in consequence, there is less danger of contamination during the critical initial stages of the fermentation. Glass jars have one further great advantage, that there is no possibility of a leakage, as so often happens with casks of small capacity when they are kept at a temperature of 65-70°F. Stoneware jars are too cold; a yeast will not thrive as it should when in contact with stone – at any rate in jars of small capacity.

Sparkling mead

A sparkling wine may be produced by bottling the mead before the after-fermentation has set in. All high-class sparkling wines are made thus; and mead will make a very fine beverage, comparable to the best sweet champagne, when treated this way.

I find that the most satisfactory results follow a bottling in February or March of a mead prepared in October the previous year. For a sparkling mead the honey-must is diluted to a specific gravity of 1.058. The primary fermentation is completed in the late autumn. As soon as the liquor has cleared, the cask should be held in a temperature below 60°F to delay the after-fermentation. Then in February or early March the clear mead is siphoned into bottles. It should not be drawn off the bottom of the cask, for some sediment would thus be inevitably drawn into the bottles. Champagne bottles should be used as the ordinary type of wine bottles are not strong enough to withstand the pressure formed by the carbon dioxide produced. It is quite a common thing for bottles to burst if the must has been made to a higher specific gravity than 1.058. As a matter of fact, a sparkling mead should not effervesce unduly; it should subside immediately on being poured into a glass, and the effervescence should be only just perceptible on the palate – it is then at its best.

To ensure a regular temperature, and thus a perfect after-fermentation, the bottles should be stored on their side in sand in a dry cellar. The corks used should be of the special champagne type and firmly wired down. Plastic stoppers, now widely used for sparkling wines, will serve the purpose equally well.

A sparkling mead is ready for use about three to four months after bottling, but it will mellow with age the longer it is kept. To ensure complete clarity, the bottles ought to be taken out of the sand about two weeks before they are used and allowed to stand upright. This is to let the slight sediment, formed in the last fermentation, settle to the bottom before the mead is brought to the table. It is advisable also to put the bottle in ice, or in a refrigerator, for about two hours before serving. A sparkling mead is greatly improved by cooling. Indeed, any medium sweet or dry mead is at its best when so treated before it is served.

The foregoing instructions on sparkling mead were set down more than twenty years ago, but still hold good. We have, however, meanwhile, found that by the adoption of an alternative procedure a far superior sparkling mead can be produced. A cask of mead is made to the instructions just outlined, but allowed to mature in wood for a full two years, then blended with a second lot prepared the previous autumn. To simplify the blending, the champagne bottles are first half filled with the two-year-old mead and thereafter topped up with mead made the previous autumn. The latter will in due course give rise to the required after-fermentation and produce the desired effervescence, but in combination with the two-year-old mead will result in an incomparably superior sparkling mead.

Sack mead

Sack mead is a sweet wine similar to sherry. It can be made in various degrees of sweetness – depending on the amount of honey to each gallon of water, and also to some extent on the age of the mead. For a medium-sweet wine I would recommend 3½ lb.-4½ lb.; for a full-bodied dessert wine 5 lb.-6 lb. But 6 lb. honey to the gallon of water is the maximum amount that should ever be used. If this amount is exceeded, the liquor will be too sweet to permit a satisfactory development of the fermentative process. The resultant product will not be mead, but merely a partially fermented syrup. Anything up to this is satisfactory, but a reading as low as 1.075 gives a rather less full-bodied mead, and one that tends to get drier with age.

The richest sack mead we make is from a must giving a reading of 1.120. This has a sugar content of 27.98 per cent or a little less than 6 lb. honey to a gallon of water. But the Maury grape wine yeast, which we use for every mead, is a particularly vigorous species and yields a rather higher alcohol content than the common sorts of yeast. This type of sack mead needs at least seven years to mature, and a mead of this quality will keep indefinitely. When fully matured it is a wondrously soft, mellow, full-bodied wine – and a wine that in its incomparable aroma holds imprisoned the subtle scent of 1,000 blossoms of moorland and countryside.

Metheglin

Sack metheglin is a spiced mead. It is made from substantially the same must as a sack mead.

A great variety of herbs and spices is often recommended in different recipes. One includes thyme, rosemary, sweet briar and other herbs; others insist on such ingredients as bitter almonds, ginger, elder, hops and rum (of all things); and other various spices such as cloves, mace, nutmeg and cinnamon. I believe most of these ingredients are much better omitted.

The peel of lemon may be added to heighten the natural flavour of a sack mead – and also of a sparkling mead. The flavour of honey and mead blend well with lemon, but the blending must be done with discretion. The peel should so merge with the flavour and aroma of mead that the addition of the peel should not be discernible as such. I recommend the peel of half a lemon to every gallon of must. The grated peel should be put into the cask just before the hot sterilised liquor is poured in. It should not be added to the must before, as, in the sterilisation, the volatile oils would be dissipated and lost.

According to our experience the only spices which seem to blend well with the natural flavour of sack mead are cloves and cinnamon – and in this instance I have in mind a very special type of honey-must, namely one made of ling-heather honey left imprisoned in the wax after the pressing. This honey is retrieved from the wax in the following way: the slabs of crushed honeycomb are melted in a bath of clean rain water, but the water should on no account be brought to the boil. Immediately the wax

121

is melted, the honey solution is drawn off at the bottom; or, alternatively, the wax is dipped off from the top into another container. The honey that was imprisoned in the wax is, of course, now retained in the water. The liquor thus obtained is allowed to cool, and then filtered according to the method already detailed. After filtering, the liquor should be crystal-clear and of a rich red-brown colour. It is diluted to a specific gravity of 1.120, and then re-sterilised, and any scum that may rise is removed.

This special type of honey-must is greatly improved by the addition of ¼ oz. cloves and 1½ oz. bark of cinnamon to every ten gallons. The spices are placed in the cask, like the lemon peel.

The sack metheglin thus obtained when fully matured will rival any brown sweet sherry in the world. It is a distinct type of sack mead unlike any other mead or wine I know. It has a character all of its own. In colour it is a rich red-brown, and the crystal-clear liquid reflects every light with an intense glow, and from it rises an evanescent perfume that is totally absent in any wine made of the juice of the grape.

CPSIA information can be obtained
at www.ICGtesting.com
Printed in the USA
BVHW061327100220
571912BV00008B/375